# Blockchain Technologies

**Series Editors**

Dhananjay Singh⊚, Department of Electronics Engineering, Hankuk University of Foreign Studies, Yongin-si, Korea (Republic of)

Jong-Hoon Kim, Kent State University, Kent, OH, USA

Madhusudan Singh⊚, Endicott College of International Studies, Woosong University, Daejeon, Korea (Republic of)

This book series aims to provide details of blockchain implementation in technology and interdisciplinary fields such as Medical Science, Applied Mathematics, Environmental Science, Business Management, and Computer Science. It covers an in-depth knowledge of blockchain technology for advance and emerging future technologies. It focuses on the Magnitude: scope, scale & frequency, Risk: security, reliability trust, and accuracy, Time: latency & timelines, utilization and implementation details of blockchain technologies. While Bitcoin and cryptocurrency might have been the first widely known uses of blockchain technology, but today, it has far many applications. In fact, blockchain is revolutionizing almost every industry. Blockchain has emerged as a disruptive technology, which has not only laid the foundation for all crypto-currencies, but also provides beneficial solutions in other fields of technologies. The features of blockchain technology include decentralized and distributed secure ledgers, recording transactions across a peer-to-peer network, creating the potential to remove unintended errors by providing transparency as well as accountability. This could affect not only the finance technology (crypto-currencies) sector, but also other fields such as:

Crypto-economics Blockchain
Enterprise Blockchain
Blockchain Travel Industry
Embedded Privacy Blockchain
Blockchain Industry 4.0
Blockchain Smart Cities,
Blockchain Future technologies,
Blockchain Fake news Detection,
Blockchain Technology and It's Future Applications
Implications of Blockchain technology
Blockchain Privacy
Blockchain Mining and Use cases
Blockchain Network Applications
Blockchain Smart Contract
Blockchain Architecture
Blockchain Business Models
Blockchain Consensus
Bitcoin and Crypto currencies, and related fields

The initiatives in which the technology is used to distribute and trace the communication start point, provide and manage privacy, and create trustworthy environment, are just a few examples of the utility of blockchain technology, which also highlight the risks, such as privacy protection. Opinion on the utility of blockchain technology has a mixed conception. Some are enthusiastic; others believe that it is merely hyped. Blockchain has also entered the sphere of humanitarian and development aids e.g. supply chain management, digital identity, smart contracts and many more. This book series provides clear concepts and applications of Blockchain technology and invites experts from research centers, academia, industry and government to contribute to it.

If you are interested in contributing to this series, please contact msingh@endicott.ac.kr OR loyola.dsilva@springer.com

Malaya Dutta Borah · Pushpa Singh ·
Ganesh Chandra Deka

Editors

# AI and Blockchain Technology in 6G Wireless Network

 Springer

*Editors*
Malaya Dutta Borah
National Institute Of Technology Silchar
Silchar, Assam, India

Pushpa Singh
KIET Group of Institutions
Delhi-NCR
Ghaziabad, India

Ganesh Chandra Deka
Ministry of Skill Development
and Entrepreneurship
Regional Directorate of Skill Development
and Entrepreneurship
Guwahati, Assam, India

ISSN 2661-8338 ISSN 2661-8346 (electronic)
Blockchain Technologies
ISBN 978-981-19-2870-3 ISBN 978-981-19-2868-0 (eBook)
https://doi.org/10.1007/978-981-19-2868-0

This Springer imprint is published by the registered company Springer Nature Singapore Pte Ltd.
The registered company address is: 152 Beach Road, #21-01/04 Gateway East, Singapore 189721,
Singapore

# Contents

# Technologies Assisting the Paradigm Shift from 5G to 6G

**Murari Kumar Singh, Rajnesh Singh, Narendra Singh, and Chandra Shekhar Yadav**

**Abstract** Artificial Intelligence (A.I.), Blockchain, and Internet of Thing (IoT)-based applications require a massive resource, enormously high data rates, and increased throughput of 1 Tbps. To accomplish these requirements, existing 5G network may not be proficient. 5G will likely not have the ability to fully meet the needs of new applications based on self-driven vehicles, drone technologies, virtual reality (V.R.), augmented reality (A.R.), mixed reality (M.R.), and extended reality (E.R.). 6G networks will be the data superhighway and make the intelligent cyber-physical world fully connected. The 6G technologies are surprisingly helpful for integrating people, devices, vehicles, sensors, information, cloud resources, and robot specialists. 6G network enables great data rate, ultra-low latency, high bandwidth speeds, high quality of services, energy efficiency, and intelligent spectrum utilization. This chapter focuses on the evolution of 1G-6G and technological advancement that enabled a paradigm shift from 5G to 6G. Performance metrics of 6G is compared with 5G, and key features of 6G are also highlighted in the proposed chapter. Terahertz communication, MIMO, and fast full-inclusion access will bring about the densification of the 6G network. Further, the chapter also introduces 6G applications with emerging technologies.

M. K. Singh (✉)
Department of Computer Science and Engineering, Sharda University, Greater Noida, U.P., India
e-mail: mksinghjamia@gmail.com

R. Singh
Department of Information Technology, GL Bajaj Institute of Technology and Management, Greater Noida, U.P., India
e-mail: rajneshcdac.mtech@gmail.com

N. Singh
Department of Management Studies, Greater Noida, GL Bajaj Institute of Technology and Management, Greater Noida, U.P., India
e-mail: narendra.naman09@gmail.com

C. S. Yadav
Department of Computer Science and Engineering, Greater Noida, Noida Institute of Engineering and Technology, Greater Noida, India
e-mail: csyadavrp@gmail.com

© The Author(s), under exclusive license to Springer Nature Singapore Pte Ltd. 2022
M. Dutta Borah et al. (eds.), *AI and Blockchain Technology in 6G Wireless Network*,
Blockchain Technologies, https://doi.org/10.1007/978-981-19-2868-0_1

**Keywords** 6G · 5G · Network architecture · Terahertz (THz) communication ·
Blockchain · Security · IoE

# 1 Introduction

The 5G network system was designed to cater for the Internet of Everything (IoE)
[1]. The 5G systems were operating at high-frequency millimeter-wave frequencies.
The 5G systems can cope with basic IoE and ultra-reliable low latency communications, but emerging technologies of IoE that demand end-to-end co-design of
communication, control, and computing localization and Akyildiz sensors remain a
challenging area that might not fulfil. AI, Blockchain, and IoE-based applications
need a tremendous amount of resources, extremely fast data rates, ultra-low latency,
and throughput of 1 Tbps. Moreover, there is a need for the best connected heterogeneous network environment [2] to serve the heterogeneous application requirements
of the user and system [3].

To overawe these encounters, the 6G network which is designed to meet the
demands of IoE applications came into existence. The 5G system could not capture
sensory inputs of Extended Reality (X.R.) because they could not deliver low latency
for high-capacity data rates. The emergence of drones, underwater communications,
and autonomous cars is a field where 5G could not meet the rate-reliability-latency
ratio. The cryptographical techniques used with 5G may not be able to secure the
IoE devices. Past systems used cellular base stations, but smart cities now require
electromagnetic surfaces such as roads and buildings to cater to wireless networks.

6G will play a critical role in transforming the IoT into related intelligent IoE,
Internet of Bodies, and Internet of Nano-things [4]. To achieve increased network
density, it will support extraordinarily high information rates of up to 1Tbps, a
very high frequency of up to terahertz (1 to 3 THz) [5], and more energy products to work on the liveness of IoT devices without batteries. 6G will necessitate
the creation of administration kinds beyond those offered by 5G. It will support
three new services: Computation-Oriented Communications (COC), Contextually
Agile eMBB Communications (CAeC), and Event-Defined uRLLC (EDuRLLC)
[6]. It will likewise operate with the arrangement of new technology, for example,
gigantic dazzling surfaces, extremely vast range receiving antennas, obvious light
interchanges, and so on.

In this chapter, author proposed a comparison of 5G and 6G with performance
metrics of 6G. The author presented the primary trend, architecture, characteristics, heterogeneous aspect of the 6G wireless network, and applications of 6G with
emerging technology. The paper is organized into five sections. Section 2 provides
background and development of related works on performance analysis on 1 G to 6
G communication systems. Section 3 describes performance measurements for 6G
networks and contrasted and regular 5G. Section 4 describes the technical characteristics of the 6G wireless network and the advantages of 6G mobile communication.

Section 5 describes the heterogeneous network aspect of the 6G wireless network. Section 6 provides the application of 6G with Emerging Technologies, and Sect. 7 concludes the paper.

## 2 Background and Evolution of Mobile Communication System

Mobile communication and network technology have started in the late 1970s and evolved noticeably with successive generations from 1G to 6G, as represented by Fig. 1. The evolution and revolution of wireless and mobile technology have reached 6G to provide a high-speed network, low latency, and integration of emerging technology.

**1G**: Voice calling was the most fundamental assistance, the voice quality was acceptable, and the networks were often all around covered. Fax services were accessible, basically to some extent, and acoustic modems took into consideration a few information transmissions. However, wireless standards were comprised of a large number of restrictive, contrary principles. In 1G (Fig. 1), the Advanced Mobile Phone System (AMPS), in view of simple balance, was set up in the United States, yet a few European nations took on the Nordic Mobile Radio (NMR) System. In Germany, the wireless system was known as the C-network. At that point, the transmission pace of AMPS was 2.4 kbps, channel capacity 30 kHz, and frequency division multiple access (FDMA) procedure were utilized for voice service. It had issues, for example, low-range use, restricted help types, and slow information service speed, among others. Roaming was unrealistic, and availability was exceptionally restricted. Therefore, just around 10% of the total populace utilized such services.

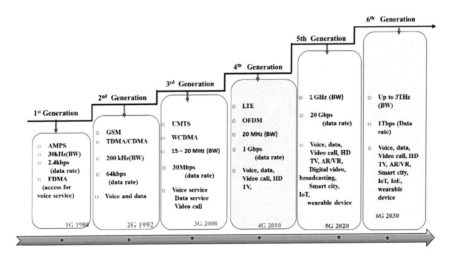

**Fig. 1** Technical specifications of different generations

**2G**: Second-generation mobile communication depends on computerized technology; the organizations like NOKIA had presented this technology dependent on GSM (Global System for Mobile Communication) standard. Finland monetarily dispatched it in 1991; it utilizes computerized regulation methods like Time Division Multiple Access (TDMA) and Code Division Multiple Access (CDMA), 64 kbps data rate and transmission capacity up to 200 kHz, as shown in Fig. 1. 2G supports global roaming with better quality and information services.

**3G**: A 3G mobile system was characterized by an association called the Third-Generation Partnership Project (3GPP). WCDMA, CDMA2000, and TD-SCDMA wireless are still up in the air to be 3G standards by the ITU (International Telecommunication Union). The first Universal Mobile Telecommunication System (UMTS) PDAs show up. It was presented in 2000, with the objective of presenting up to 2 Mbps data rate appropriate for rapid speed Internet, 15 to 20 MHz data transmission. It could give sight and sound services like sound, video, pictures, messages, and web perusing. It uses WCDMA (Wireless Code Division Multiple Access) technology for UMTS. Due to similarity issues among various standards (like WCDMA, CDMA2000, and Wi-Max) in UMTS, self-impedance issue in CDMA, complected design of WCDMA, and sending cost was high. It was hard to address the issue of clients perusing the information for minimal price and **rapid** [7].

**4G**: Fourth era mobile networks were presented in the late 2010s, and the objective of 4G was to offer top-notch types of assistance like high speed, high capacity, better security, and minimal expense services for information and voice. In this technology, all mobile devices are associated through I.P.-based network system. 4G utilizes the OFDM procedures, which overcome the essential deficiency of CDMA strategy utilized in 3G. Moreover, it fuses some more wireless technology like LAS-CDMA, MC-CDMA, and organization LMDS to give smooth wandering from one technology to another. However, Long-haul Evolution (LTE) and Wi-Max are moreover considered as 4G technology. Hypothetically, it could present up to 1 Gbps data rate with 20 MHz data capacity. Moreover, the data rate of this innovation had met the fundamental prerequisite of clients, however because of novel and arising technology applications, similar to AR/VR, IoT, UAV (Unmanned Arial Vehicle), and so forth required more transfer speed and data rate required.

**5G**: The standard archive of 5G had been distributed by 3GPP authoritatively endorsed in the Plenary Meeting in June 2018. It incorporates ultra-Reliable Low-Latency Communication (uRLLC), massive Machine-Type Communication (mMTC), and enhanced Mobile Broadband (eMBB) administration. URLLC upholds unwavering high quality and low inertness (in millisecond), eMBB upholds super high throughput to address the need of clients getting to the original application, AR/VR, video real time, and so forth. The objective of 5G is to accomplish the communication between machine to machine(M2M), the interconnection between individuals and machines and give rapid, full inclusion, limitless admittance to information, capacity to share data whenever anyplace, and intelligence network.

The center technology of 5G incorporates an enormous scope of MIMO radio wire at the actual layer to help higher transmission capacity. Albeit the fifth era support ultra-reliable low-latency communication, it has the restriction of the short-packet, recognizing-based uRLLC limits that limit the movement of low-latency services with higher data rate, similar to AR/VR, blended reality, arising Internet of Things, and so forth.

## 2.1   Limitations of 5G: Driving Force to Move Toward 6G

Despite the way that 5G communication networks can help URLLC, the short-packet, distinguishing-based URLLC components of 5G infer that there are impediments to how well 5G can offer sorts of help that require low-latency and high-reliability quality while ensuring high data rates. In the future, Industry 4.0 and Industry 5.0 will have profoundly thick M2M (machine-to-machine communication), smart vehicles, and driverless vehicles will require ultrabroad and real-time services, like AR/VR cloud gaming, 8 K/4 K video, and 3D holographic video. A smart portable robot will require superior registering assets through remote access.

Toward 2030, UAV, AGV (automated guided vehicle), and Swarm Networking (S.N.) will be normal for the associated calculated and transport, and crisis reaction, and so forth. They will request continuous administrations with hearty connectivity [8]. The solicitations of low latency with high data rates, communication and identifying gathering, and open interfaces require the first architecture engineering design of 6G [9]. Subsequently, for example, the multisensory X.R. applications (VR/ AR/MR) that need such extraordinary necessities can't arrive at their maximum capacity in a 5G cellular network. Additionally, the intermingling of capacities like communication, control, processing, and detecting is a prerequisite for impending applications of the Internet of Everything (IoE). This angle has been generally disregarded in 5G. All together for the fruitful activity of IoE services like independent system, the organization needs to furnish heterogeneous devices with high data rates and dependability, low latency for both uplink and downlink exchanges [10]. To move beyond these downsides, a problematic 6G wireless organization is required that has been planned while remembering the prerequisites of such applications and all related patterns in expertise that may emerge from these arising IoE facilities.

**6G**: The 6G network will guarantee that all that will be associated profoundly, wisely, and seamlessly. Based ashore, space, air, and ocean, the 6G organization will utilize high-throughput satellites to establish circle satellite communication, planes communication, acoustic communication, and so forth [11]. 6G will actually want to elevate satellites to meander and to shift between whichever two satellites. It will incorporate different stages for marine Internet communication (in light of the ocean) and ground mobile communication (in view of the land) to shape a high-accuracy, high-dependability network covering the whole world. Numerous 6G-related technology is being created from various points [8].

Addresses the development of basic and massive machine-type communication (MTC) toward 6G [12, 12]. However, the cultural turn of events and use-cases relevant to machine-type communication toward 2030; distinguish key execution pointers (KPI) and prerequisites; and examine various expected specialized empowering influences for MTC in future 6G networks. M. Katz et.al. present the underlying vision of what is 6G, the creators have introduced the examination plan of the as of late made thought 6GFP (6 Genesis Flagship Program). The 6GFP is an 8-year research drive of a consortium shaped by driving Finnish scholastic and modern accomplices, targeting creating, executing, and testing key empowering innovations for 6G [13]. David and Berndt [14] have presented a vision of the 6G system and its requirements, such as energy collection tactics, remote charging for battery life, and the use of artificial intelligence to collaborate with creative administrations. They've identified certain critical needs that 1G through 5G won't be able to meet. Among them are ultralong (preferably limitless) battery life and a lack of concentration on high data rates (like in 5G), both of which are only available in single islands. In the papers [15–17], creators have further elaborated the principal measurements, design driver variables of 6G. P. E. Mogensen characterizes the six key measurements to plan 6G, creators have distinguished the accompanying key innovation changes as having the most elevated potential to characterize for the 6G network: (i) A.I./ML-driven air interface plan and advancement; (ii) venture into novel choice groups and new intellectual range sharing techniques; (iii) the integration of limitation and detecting abilities into framework definition; (iv) the accomplishment of outrageous execution necessities on latency and unwavering quality; (v) fresh network design ideal models including sub-networks and RAN-Core intermingling; and (vi) new security and protection plans [17, 18]. The 6G network looked at the vital provisions of Intelligent Radio (I.R.), software-defined radio (SDR), and cognitive radio (C.R.). They have endeavored to give a forward-looking examination guide for 6G [15]. In [16], authors have researched the driver factors for 6G (i.e., holographic, massive availability, and time delicate/time designed applications), the 6G networks plan standards, and engendering qualities of the 6G organizations.

## 3  Performance Metrics for 6G Networks and Compared with Conventional 5G

Differentiated and 5G, 6G will have a high peak rate, a high customer experience rate, low latency, high affiliation density, strong mobility, high traffic density, high faithful quality, high network energy efficiency, high reach effectiveness, strong situating ability, strong range support capacity, and so on. An examination is portrayed in the accompanying Fig. 2.

The absolute most expected scenarios of uses like eMBB-Plus, Big communication (BigCom), Secure super ultra-reliable low-latency communications (SURLLC), Three-dimensional incorporated interchanges (3D-InteCom), Unconventional Data Communications (UCDC), holographic communication, tactile interchanges, human-security communications that will be upheld by 6G communication [19, 20]. These applications lead to recognize the key provisions that are needed in

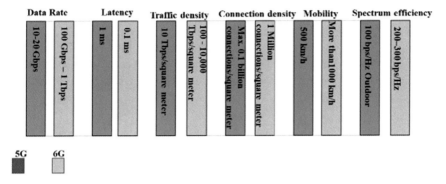

**Fig. 2** Relative performance comparison between 5 and 6G

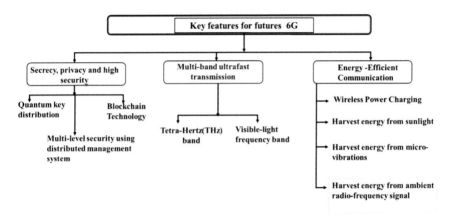

**Fig. 3** E.M. (electromagnetic) spectrum and wavelength of THz and mm waves

such communication. Yastrebova et.al. have anticipated various modern, still non-existent wireless communication conditions for the 2030s, including holographic calls and a material Internet [21]

Figure 3 shows the key features (high security, multi-band ultrafast transmission, and energy productive correspondence) for future 6G.

## 3.1 High Security, Privacy, and Secrecy

Conventional encryption methods dependent on the cryptosystem like RSA (Rivest Shamir–Adleman) public-key algorithm are as yet utilized in 5G network to give transmission security and privacy [21]. Under the interest of Big Data and A.I. technology, RSA cryptosystems have become unstable, yet original security assurance strategies are as yet far off in the 5G future. Physical security and quantum key dissemination through VLC ought to be utilized to address information security challenges in 6G system. For a far-reaching guarded technique against digital assaults, complex quantum figuring and quantum specialized strategies might demonstrate useful [22–24]. Accomplishing absolute anonymization, decentralized security, non-disseminated, and untraceability can be accomplished through the utilization of blockchain technology [25].

## 3.2 Multi-Band Ultrafast Transmission

The information hungry applications such as M2M, smart vehicles, cloud gaming, driverless vehicles, AR/VR, 3D holographic video 8 K/4 K video, UAV, AGV, and Swarm Networking need an extremely high-range data transmission that is at present inaccessible in the millimeter-wave (mm-wave) range. The present circumstance has genuine ramifications for the region or ghostly spatial productivity, just as the recurrence range groups are essential for connection. Thus, a more extensive radio recurrence range data transmission is required, which must be found in the sub-THz and THz groups (displayed in Fig. 4), generally known as the hole band among microwave and optical spectra [26].

For high-velocity correspondences, the text has introduced a brief and thorough investigation of the THz band in the 0.1–10 THz range [27]. As an aftereffect of late developments in microwave communications, THz electrical and photonic advances are presently conceivable. 6G hybrid handsets will doubtlessly utilize a THz/free-space-optical hybrid. A THz signal can be produced by the optical laser, or an optical signal can be sent [28].

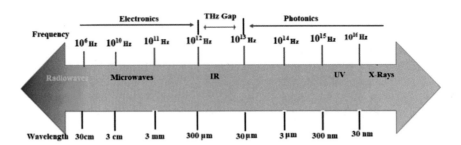

**Fig. 4** E.M. (Electromagnetic) spectrum and wavelength of THz and mm waves

# 4 The Technological Characteristics of 6G Wireless Network and Advantages of 6G Mobile Communication

6G mobile communication technology has mechanical advances like universal inclusion, security, and AI-inserted self-intelligence [18, 29].

**Ubiquitous coverage**: 6G has the capacity of incorporating air, space, ocean, and land, eliminating signal vulnerable sides, and working with clients to get to a smooth and far-reaching network. For instance, to tackle the issue of inadequate inclusion of existing earthbound communication systems, 6G will acquaint air-with space communication networks (e.g., satellite communication networks) to further develop inclusion. In any case, the satellite communication network has the issue of lacking limit. It will be important to consolidate the huge limit of the ground communication system with the high inclusion of the satellite communication network to shape a crossbreed organization. Likewise, 6G will empower individuals and things to get to the organization whenever and any place. Moreover, the performance of wireless mobile technology is analyzed to increase the performance of infrastructureless networks in terms of packet loss, end-to-end delay, packet loss, node density, routing overhead, number of acknowledgements, and node energy [30–32]. They can rapidly partake in information trade without being tormented by constant system heterogeneity. System development and network speed can coordinate with the lean turn of events.

**Security**: 6G performs start to finish prejudgment and characterization of information security. It upholds client characterized methodologies to guarantee the security of terminal nodes, air interfaces, information transmission, and client discernment. Simultaneously, 6G doesn't have to meet the basic communication security prerequisites yet gives separated security services to various situations. Henceforth, 6G can adjust to an assortment of conventional or new network access draws near, ensure client protection, and uphold open and dispatchable security abilities. In contrast to the past ages, 6G incorporates both acknowledging and virtualizing resources. The security prerequisites that relate to the physical and virtual conditions are unique. This advances the requirement for higher necessities for security arranging, strategy making, and algorithm design. As it understands and virtualizes network resources (and, it ought to be noted, both physical and virtual assets will expand network heterogeneity), 6G will manage both physical and virtual security necessities.

**AI-embedded self-intelligence**: 6G will be joined with A.I. to achieve mindfulness, self-learning, self-enhancement, and self-development. Taking the design, activity, and streamlining of a 6G system dependent on A.I., for instance, the benefits can be found in the accompanying situations: a client experience-driven network setting, an adaptable and self-enhancing network, proactive network observing and disappointment investigation, and proactive network security protection [33]. Some more mechanical benefits of 6G portable correspondence relying on technological forward leaps are displayed in Fig. 5.

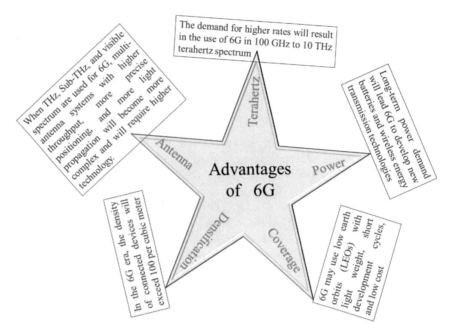

**Fig. 5** The technological advantages of 6G mobile communication

## 4.1 Concept of Terahertz Communication

As indicated by [34, 35], sixth-generation communication technology arises around 2030. 6G requires a bigger transmission capacity to help high information rates and ultrafast communication. These particulars lead to the foster Terahertz range framework, which comprises the recurrence band of 0.1 to 10THz [36–38]. All in all, Terahertz (THz) is a recurrence range in the far-infrared range, i.e., at the interface of hardware and photonics. The lower limit is directly over the microwave zone, which is utilized by satellite dishes and cell phones, and as far as possible is close to the infrared frequencies utilized in T.V. controllers and other devices. As an outcome, the THz hole consolidates microwave and infrared electromagnetic range. This locale of the electromagnetic range is wealthy in detecting, imaging, and correspondences prospects, just as one of a kind uses in evaluating for weapons, explosives, and biohazards, just as water content and human skin.

The primary benefit of the terahertz antenna is to get exceptionally enormous transfer speed, just as the most straightforward adjustment plans. This can be utilized for different applications like security, research, adaptable electronic devices, far off discovery, detecting and picture applications, energy protection devices, and specialized gadgets. It has a potential for very high information rates, for example, 1 Terabit each second [40], which is a significant necessity of a 6G communication network.

At first, it has been utilized essentially for space and stargazing applications. New application areas found in medication, the natural, and organic sciences, just as in security and quality control, have arisen as of late [41, 42]. Because of less free space diffraction of the waves, terahertz communication is more directional than microwave or millimeter links [43]. It can uphold ultrahigh transfer speed spread range networks, which can empower secure communication enormous limit system, and insurance against channel jamming attacks. There is a lower weakening of THz radiation contrasted with I.R. under certain climatic conditions, e.g., haze. THz can empower solid communication where IR-based networks would come up short under certain climate conditions and for explicit link length necessities.

Interestingly, with 5G, 6G expected to help higher data rate or capacity (1 terabit each second), higher intermingling, lower latency, higher reliability ($10^{-9}$), lower energy and cost, high density [44], overall networks, free battery IoT devices, associated intelligence with A.I. [12], and another blend of these prerequisites for impending use-cases.

Since their extents are near the THz band's frequency, atoms and infinitesimal particles in the climate, like water fume and oxygen, assimilate or disperse the THz recurrence band. Since way loss is very high for 6G, it forces a genuine constraint on communication reach and inclusion region which is up to 10 m. So essential necessity is to cultivate radio antennas which resound in THz locale and supports high data transfer capacity just as exceptionally mandate. Graphene sheet-based antenna is one of the fields which required further exploration. Channel and commotion displaying, actual layer, connect layer, network layer, and transport layer are, for the most part, issues in THz band communication networks. Physical layer difficulties incorporate balance plans, channel coding plans, different info and numerous yield (MIMO) networks, transfers, media access controlling (MAC) [45], and synchronization. However, the antenna needed for 6G correspondence framework ought to uphold high information rate, yet it ought to be a multi-band resounding radio wire. Relies on the applications, now and then, an omnidirectional antenna is required, while a profoundly directional antenna may likewise need for explicit purposes. The size ought to be exceptionally minimal, creation cost ought to be extremely low and ideally, it ought to be circularly captivated polarized antenna.

Multiple-Input Multiple-Output (MIMO) antenna cluster innovation might be utilized for 6G reasons since it might give high addition and reciprocally upgrade the correspondence reach and limit of the organization. Many examinations are as yet proceeding to foster some maritime antenna [46] fostered a funnel-shaped horn antenna that works at 0.270–0.330 THz which is created with wire-cutting EDM technology, displayed in Fig. 6.

The 16-component planar exhibit of depression supported fix antenna is created by [47] utilizing multi-facet high-frequency PCB technology, which reverberates between 0.135 and 0.115 THz as shown in Fig. 3.

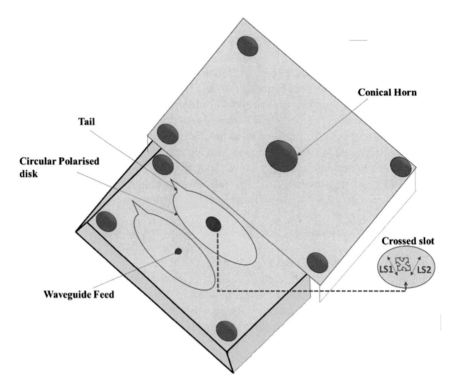

**Fig. 6** Conical horn antenna 3D exploded view

## 4.2  *Densification*

Fast full-inclusion access will bring about the densification of the 6G network. The thickness of associated devices will surpass 100 for every cubic meter. 6G base stations will address the issue of getting to a great many remote devices simultaneously; subsequently, new access components will be expected to deal with a lot of client information in a proficient and versatile way. Because of the densification of 6G system, the entrance capacity in excess of 100 devices for each cubic meter has carried new applications to shrewd homes, keen urban communities, fiasco counteraction, public well-being and protection, Medicare, and education. 6G will provide communication among individuals and things, or among things and things [48].

## 4.3  *Power*

The drawn-out power request will lead 6G to develop new batteries and remote energy transmission innovations. The coming of 6G will advance examination on new battery technologies, like strong state batteries and graphene batteries. Batteries in 6G will have higher energy, lower cost, more significant amp-hour storage capacity, and more convenient capacities. Utilizing remote energy transmission to charge 6G batteries is relied upon to broaden battery life and make charging more helpful [49].

## 5  Heterogeneous Network Aspect of the 6G Wireless Network

6G networks will rely heavily on network intelligence, with the network performing dynamic activities in response to changing environmental conditions. According to experts, artificial intelligence will play a critical role in the Internet of Things (IoT) and Internet of Bio-Networks (IoBNTs)-driven future. The most efficient technique to transport data will be the key to moving from 5 to 6G. A system that does not require human intervention will be perfect [46, 50].

Figure 7a shows a comparison between 5 and 6G network architectures. High processing power, intelligence, and high capability have updated the 6G core network to the 5G core network. Incorporating Base Stations (B.S.), satellites and unmanned aerial vehicles (UAVs) will also improve the access network. In 6G, there is a vertical hand besides the horizontal hand found in 5G. In addition, fog computing and Multi-access Edge Computing (MEC) are essential elements in 6G network design, which reduces the need of a great number of devices in the user plan by latency and bandwidth for widely used services. Clouds, Edge, and Fog (CEF) computing are used to provide quick access to services. Virtualization, softwarization, and slicing will be used to accomplish the properties of self-organization, self-optimization, and self-reconfiguration.

## 5.1  Cloudification/Fog/Edge

There are thousands of sensors in companies, and hundreds of sensors in homes are installed to provide a significant amount of data and it is extremely difficult to wire all of these sensors together. Furthermore, these equipment are sophisticated and intelligent, capable of judiciously evaluating and using minimal computer resources. Thus, the data must be downloaded from the cloud to the end of the gadget. To reduce the processing latency from a cloud perspective, we require to move the processing nearer to terminal devices. To enhance the service quality, we need to shift the burden closer to the edge.

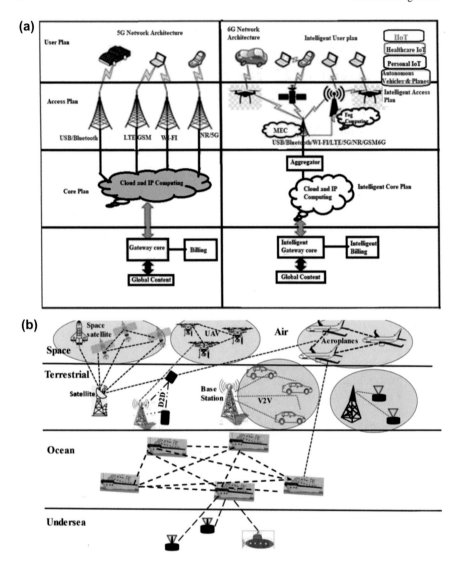

**Fig. 7** **a** A comparative analysis between 5 and 6G network architecture **b** Heterogenous network architecture of 6G

## 5.2   Softwarization

Configurability, programmability, self-organization, flexibility, and heterogeneous use-cases are the primary driving forces behind the creation of 6G networks. The physical equipment that enables all of the aforementioned capabilities is tough to install. Virtualization and softwarization are the two utmost challenging hypotheses for 6G networks since they enable underlying networks to realize functions. Softwarization is established of boundaries and conventions that allow the control and user planes to be separated and the network to be established in software. The user plan is often made up of a collection of distributed routing tables and stateless tables, which execute high rates for packet switches. The centralized control plan, which gives information on end-to-end routing for different services, updates these tables. The service provider and SDN provider are in charge of data interchange and administration. The needed service will be sent to the service client by SDN providers at the end. The service client can control these services by using these virtual resources.

## 5.3 Virtualization

Virtualization of the network function permits the implementation of software functions on virtual machines while still providing access to common physical resources, such as storage, networking, and computation. Within the same virtual machine, several functions with containers can be created. Continuous virtual machines can handle continuously varying network needs, such as network traffic and services. The virtualized services include (i) service baseband processing, (ii) load management, (iii) mobility management, and (iv) evolved packet core (EPC).

## 5.4 Slicing

One of the center networks is included that will permit us to build an adaptable network on top of an actual common infrastructure which is network cutting. Moreover, 5G comes to shape, it will end up being a center technology for empowering a wide scope of utilizations. Utilizing a single piece of network equipment to provide several logical networks across a shared physical infrastructure at a lower cost. Slices can be given to certain use-cases such as IoTs, service classes, consumer classes, and wireless network operators. It can also be attributed to other types of networks, such as wire versus wireless or consumer versus commercial. To configure a new slice is the most challenging job in network slicing since it impacts the complete network components. We may also construct slices for health care, automotive, utilities, etc.

The 6G network, which includes undersea, ocean, and all aerial communications, is depicted in Fig. 7b. As illustrated in Fig. 7b, 6G will allow the devices to connect with extremely low data rates, such as Internet of Things and biosensors devices, and it also enables the high data rate communication, such as High-Definition (H.D.) video transmission in smart cities. In a 6G network fast-moving vehicle or jet, communication will be feasible. It also demonstrates that all of the networks will be combined. In addition, the IRS may be installed on bridges, buildings, and surfaces in smart cities to improve coverage and quality of service (QoS) for each communicating device. The reliable underwater data connections will permit connectivity among submarines, ships, and sensors at the undersea level in maritime communication situations. Furthermore, cutting-edge technology such as augmented reality (AR/VR), haptics, and machine learning (ML) will further minimize the impact of physical distances throughout the world.

## 6 Application of 6G with Emerging Technologies

This chapter explores recent results in several 6G wireless network aspects with the integration of A.I. and Blockchain Technology and evaluates the performance of technologies used in the future. A.I. and Blockchain have the potential for dynamic resource management and mobility management in 6G network. The integration of A.I. and Blockchain in 6G will enable the wireless network to monitor and manage resource utilization and sharing efficiently. Due to a limited number of resources, Blockchain users may have to compete for wireless channel to broadcast transactions using a media access control (MAC) mechanism. The A.I. techniques enabled an efficient and intelligent resource sharing mechanism. A.I. and Blockchain efficiently optimized the network performance, AI-enabled wireless edge processing, brilliant mobility of the executives, handover the board, and intelligent spectrum of the executives, and decentralized network management in the parts of 6G.

### 6.1 Application of 6G with Blockchain

Blockchain and distributed ledger technology (DLT) applications might be seen as the up-and-coming age of appropriated detecting administrations, requiring a synergistic mix of URLLC and monstrous machine-type interchanges (mMTC) to offer low latency, adaptability, and constancy. These are the most problematic IoE technologies. Zhang et al. [66] have completely considered ongoing scholastic work on utilizing blockchain to remote organizations according to alternate points of view like protection insurance, secure access control, asset sharing, management, and certificate. The creators likewise proposed a bound together engineering of Blockchain Radio Access Network (B-RAN) as a dependable and safe paradigm for 6G networks by utilizing blockchain technology with expanded proficiency and security [51]. Blockchain with 6G can revolutionize the health care and agriculture domain [52]. Sekaran et.al. have introduced the most well-known blockchain application like IoV (Internet of Vehicle), UAV, SCM, food ventures, agribusiness, smart network, smart assembling, medical services, and so on, for 6G-empowered IoT as shown in Fig. 8 [53].

The authors [54] underscored the need for 6G, just as the latest 6G research endeavors, applications, and fundamental empowering advances for 6G networks. Intelligent resource of the executives, trust-building, fast, secure and transparent

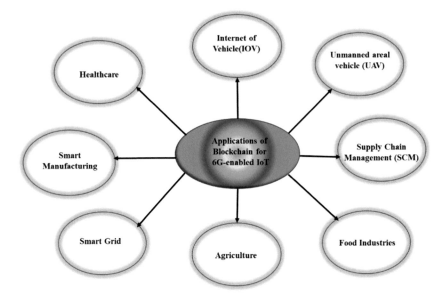

**Fig. 8** Applications of Blockchain for the 6G-enabled network

communication, cost decrease, and decentralization are the benefits of blockchain in 6G [55].

## 6.2 Application of 6G with A.I.

6G networks can provide AI-enabled applications like optimization of healthcare planning, financial market monitoring, and advance forecasting, and for a variety of mobile devices by utilizing enhanced wireless communications. In some cases, it may be able to do A.I. computations on par with a human brain and control in real time. This section will look at two important A.I. applications in the context of 6G wireless networks: scheduling of autonomous vehicles and indoor positioning [56].

### 6.2.1 Scheduling of Autonomous Vehicle

The difficulties of independent vehicle scheduling for keen urban communities are anticipated to be settled by 6G coordinated with A.I. technology. To drive securely and reliably out and about in the 6G future period, autonomous vehicles can utilize installed sensors to procure data about the nearby climate and control the bearing and speed dependent on information about a vehicle's position, obstacles, etc. Autonomous vehicles can communicate with other street clients, like walkers, cyclists, and other autonomous vehicles, as well as knowing their whereabouts, environmental elements, and street conditions. At the point when an autonomous vehicle moves toward a convergence, for instance, it can rapidly build a constant dynamic network connected with the crossing point and suspend it in the wake of passing, as shown in Fig. 9.

The data uploaded to the cloud server can subsequently be analyzed to determine a driving route and calculate travel times. Other application domains may include healthcare optimization planning, financial market monitoring, and advanced forecasting, among others.

### 6.2.2 Indoor Positioning

Indoor placement is critical for emergency rescue, safety surveillance, firefighting, and other things. The crucial stage of rescue in earthquakes, fires, and other catastrophes is to locate the trapped persons, as shown in Fig. 10 rapidly.

During an emergency rescue, it can give significant technological help and guarantee the security of trapped persons. Through data analysis, indoor situating may catch a client's movement, follow and trace the client's situation with conduct and interest preferences. It is utilized to find the constant development or position track of an individual through the scientific method. Nowadays, indoor positioning is being utilized extensively in mobile payment, indoor navigation, in-store shopping guidance, person movement analysis, item monitoring, and other human-related tasks. So it has great commercial value.

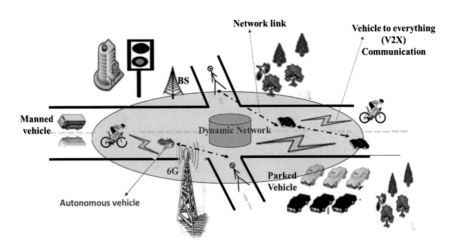

**Fig. 9** Autonomous vehicles in 6G network using A.I

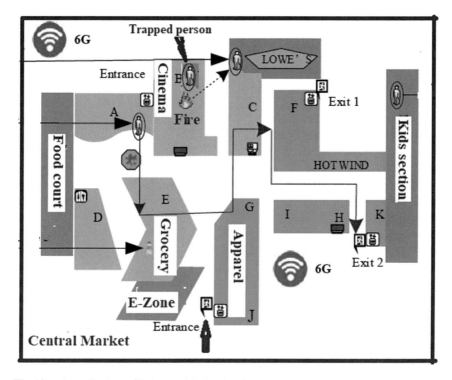

**Fig. 10** A.I. applications of indoor positioning in 6G network

**Summary and Comparisons of 1G to 6G Technology**

Table 1 provides a detailed comparisons between the main elements of 1G, 2G, 3G, 4G, 5G, and 6G architectures. It shows description about features like network characteristics, technological development, bandwidth, multiplexing, latency, modulation, web standards, application etc.

# 7 Conclusion

In this chapter, we have explored the progression of mobile generations through 1G to 6G. We initially introduced the 1G–6G system and compared the 6G and 5G performance matrices. We have also highlighted all of 6G's technological features and important key technologies. High security, privacy, secrecy, and ultra-high-speed transmission are some of the major advantages of 6G, and the use of 6G in blockchain technology is clarified. The use of terahertz frequency communication, which enables extremely high bandwidth and ultra-high-speed transmission, is also highlighted, as these are the major demands for 6G. The specifications, needs, and designs of antennas are provided, followed by a discussion of general difficulties in the THz

**Table 1** Comparisons of various characteristics among the 1G, 2G, 3G, 4G, 5G, and 6G

| Generation | 1G | 2G | 3G | 4G | 5G | 6G |
|---|---|---|---|---|---|---|
| Network Characteristics | Voice | Packet Switching for data, Circuit switching | Packet Switching, broadband | AallP,Packet switching, Ultrabroadband | Packet Switching, Cloudization, Softwarization, Virtualization, Slicing, www | Intelligence, Cloudization, Softwarization, Virtualization, Slicing, www |
| Technological Development | AMPS | GSM, GPRS, TDMA, EDGE | CDMA, WCDMA, CDMA-2000, UMTS | MIMO, OFDM, carrier aggregation, D2D communication | Cloud, Fog, EDGE, MM wave, LDPC, Massive Mimo, NOMA | THz communication, Quantum communication, Machine learning, AI, Blockchain |
| Peak data rate | 2.4 Kbps | 64 Kbps | 30 Mbps | 1 Gbps | 20 Gbps | 1 Tbps |
| Mobility | – | – | – | 350 km/h | 500 km/h | > = 1000 km/h |
| Latency | – | 300 ms | 100 ms | 10 ms | 1 ms | 10–100 microsecond |
| Multiplexing | FDMA | TDMA/CDMA | CDMA | CDMA | CDMA | OFDM |
| Modulation | FM, FSK | GMSK | QPSK | SU-MIMIO MU-MIMO TDD FDD | FQAM FBMC Massive MIMO Advanced MIMO | FQAM FBMC Massive MIMO Advanced MIMO |
| Web Standard | – | WWW | WWW(IPv4) | WWW(IPv4) | WWW(IPv6) | WWW(IPv6) |
| Location of First commerc-ialization | USA | Finland | Japan | South Korea | San Marino | Yet to be implemented |

(continued)

**Table 1** (continued)

| Generation | 1G | 2G | 3G | 4G | 5G | 6G |
|---|---|---|---|---|---|---|
| Application | Voice | VOICE, TEXT | VOICE, MULTIMEDIA | VOICE, Mobile Internet, Mobile TV, MOBILE PAY, HD VIDEOS | Virtual Reality, Augmented Reality, $360^0$ Videos, UHD videos, Wearable Device, Internet of Things, Smart City, Telemedicine | Fully automated vehicle, Driverless car, Holographic Verticals, Digital sensing, Deep-Sea Sight |

band in 6G network communication systems. The 6G network's heterogeneous architecture, which is a four-tier integrated space-air-ground-underwater network, has been introduced. Then we looked into how A.I. and blockchain may be used in the 6G network.

## References

1. Zikria YB, Kim SW, Afzal MK, Wang H, Rehmani MH (2018) 5G Mobile services and scenarios: challenges and solutions
2. Singh P, Agrawal R (2018) A customer centric best connected channel model for heterogeneous and IoT networks. J Organizat End User Comput (JOEUC) 30(4):32–50
3. Singh P, Agrawal R (2019) AHP based network selection scheme for heterogeneous network in different traffic scenarios. Int J Informat Technol 1–9
4. Zhang Z, Xiao Y, Ma Z, Xiao M, Ding Z et al (2019). 6G wireless networks: vision requirements, architecture, and key technologies. IEEE Vehicular Technology Magazine, 14(3), 28–41
5. Xu G (2020). Research on 6G mobile communication system. Journal of Physics: Conference Series, 1693(1). https://doi.org/10.1088/1742-6596/1693/1/012101
6. Letaief KB, Chen W, Shi Y, Zhang J, Zhang YJA (2019) The roadmap to 6G: A.I. empowered wireless networks. IEEE Commun Magazine 57(8), 84–90
7. Chen H-H (New York, 2007) CDMA Technologies, Wiley, Chaps. 2.6.5– 2.6.7
8. Mahmood NH, Böcker S, Munari A, Clazzer F, Moerman I, Mikhaylov K, Lopez O, Park OS, Mercier E, Bartz H, Jäntti R (2020). White paper on critical and massive machine type communication towards 6G
9. Chowdhury MZ, Shahjalal M, Ahmed S, Jang YM (2019) 6G wireless communication systems: applications, requirements, technologies, challenges, and research directions. arXiv 2019. arXiv:1909.11315v1
10. Saad W, Bennis M, Chen M (2019) Accepted from open call a vision of 6G wireless systems : applications , trends , technologies , and open research problems. IEEE Netw 1–9. https://doi.org/10.1109/MNET.001.1900287
11. Lu Y, Zheng X (2020) 6G: A survey on technologies, scenarios, challenges, and the related issues. J Ind Inf Integr 19(June):100158. https://doi.org/10.1016/j.jii.2020.100158
12. Chowdhury MZ, Shahjalal M, Ahmed S, Jang YM (2020) 6G wireless communication systems: applications, requirements, technologies, challenges, and research directions. IEEE Open J Commun Soc 1: 957–975. https://doi.org/10.1109/ojcoms.2020.3010270
13. Katz M, Matinmikko-Blue M, Latva-Aho M (2018) 6Genesis flagship program : building the bridges towards 6g-enabled wireless smart society an ecosystem. In: 2018 IEEE 10th Latin-American conference on communications (LATINCOM), 1–9
14. David K, Berndt H (2018) 6G Vision and requirements: Is there any need for beyond 5g? IEEE Veh Technol Mag 13(3):72–80. https://doi.org/10.1109/MVT.2018.2848498
15. Drivers K, Requirements C, Architectures S (2019) 6G technologies: key drivers, core requirements, system architectures, and enabling technologies. IEEE
16. Tataria H, Shafi M, Molisch AF, Dohler M, Sjoland H, Tufvesson F (2021) 6G wireless systems: vision, requirements, challenges, insights, and opportunities. Proc IEEE. https://doi.org/10.1109/JPROC.2021.3061701
17. Tataria H, Shafi M, Molisch AF, Dohler M, Sjoland H, Tufvesson F (2021) 6G wireless systems: vision, requirements, challenges, insights, and opportunities. Proc IEEE. https://doi.org/10.1109/JPROC.2021.3061701
18. Yrjola S, Ahokangas P, Matinmikko-Blue M, Jurva R, Kant V, Karppinen P, Kinnula M, Koumaras H, Rantakokko M, Ziegler V, Thakur A (2020) White paper on business of 6G (Issue 3). http://arxiv.org/abs/2005.06400

19. Ziegler V, Yrjola S (2020) 6G indicators of value and performance. In 2020 2nd 6G wireless summit (6G SUMMIT), pp 1–5. IEEE
20. Akyildiz IF, Jornet JM, Han C (2014) Terahertz band: Next frontier for wireless communications. Physical communication 12:16–32
21. Khalid M, Amin O, Ahmed S, Shihada B, Alouini MS (2019) Communication through breath: Aerosol transmission. IEEE Commun Mag 57(2):33–39
22. You X, Wang CX, Huang J, Gao X, Zhang Z, Wang M, Huang Y, Zhang C, Jiang Y, Wang J, Zhu M (2021) Towards 6G wireless communication networks: vision, enabling technologies, and new paradigm shifts. Sci China Informat Sci 64(1). https://doi.org/10.1007/s11432-020-2955-6
23. Fu Y, Shuanlin LIU, Yabin GAO, Xiuzhong CHEN (2019) U.S. Patent No. 348,493. Washington, DC: U.S. Patent and Trademark Office
24. Jha MS, Maity SK, Nirmal MK, Krishna J (2019) A survey on quantum cryptography and quantum key distribution protocols. Int. J. Adv. Res. Ideas Innov. Technol 5:144–147
25. Obeed M, Salhab AM, Alouini MS, Zummo SA (2019) On optimizing VLC networks for downlink multi-user transmission: a survey. IEEE Commun Surv Tutorials 21(3):2947–2976
26. Henry R, Herzberg A, Kate A (2018) Blockchain access privacy: Challenges and directions. IEEE Secur Priv 16(4):38–45
27. Elayan H, Amin O, Shihada B, Shubair RM, Alouini MS (2019) Terahertz band: The last piece of R.F. spectrum puzzle for communication systems. IEEE Open Journal of the Communications Society 1:1–32
28. Rappaport TS, Xing Y, Kanhere O, Ju S, Madanayake A, Mandal S, Trichopoulos GC (2019) Wireless communications and applications above 100 GHz: Opportunities and challenges for 6G and beyond. IEEE access 7:78729–78757
29. Sengupta K, Nagatsuma T, Mittleman DM (2018) Terahertz integrated electronic and hybrid electronic–photonic systems. Nature Electronics 1(12):622–635
30. Dang S, Amin O, Shihada B, Alouini M-S (2020) What should 6G be? Nat. Electron. 3:20–29
31. Singh R, Singh N (2021) Performance Analysis of TCP Newreno Over Mobility Models Using Routing Protocols in MANETs." International Journal of Wireless Networks and Broadband Technologies (IJWNBT) 10.2 (2021): 1–15
32. Singh R, Singh N (2020) Performance assessment of DSDV and AODV routing protocols in mobile adhoc networks with focus on node density and routing overhead. In: 2020 international conference on emerging smart computing and informatics (ESCI). IEEE
33. Singh R, Singh N, Dinker AG (2021) Performance analysis of TCP variants using AODV and DSDV routing protocols in MANETs. Recent advances in computer science and communications (Formerly: Recent Patents on Computer Science) 14.2 (2021): 448–455
34. Mogensen PE (2020) Communications in the 6G Era. https://doi.org/10.1109/ACCESS.2020.2981745
35. Giordani M, Polese M, Mezzavilla M, Rangan S, Zorzi M (2020) Toward 6G networks: Use cases and technologies. IEEE Commun Mag 58(3):55–61
36. Latva-aho M, Leppänen K, Clazzer F, Munari A (2020) Key drivers and research challenges for 6G ubiquitous wireless intelligence
37. Akyildiz IF, Han C, Nie S (2018) Combating the distance problem in the millimeter wave and terahertz frequency bands. IEEE Commun Mag 56(6):102–108
38. Khan LU, Yaqoob I, Imran M, Han Z, Hong CS (2020) 6G Wireless Systems: A Vision, Architectural Elements, and Future Directions. IEEE Access 8:147029–147044. https://doi.org/10.1109/ACCESS.2020.3015289
39. Tekbıyık K, Ekti AR, Kurt GK, Görçin A (2019) Terahertz band communication systems: challenges, novelties and standardization efforts. Phys Commun 35:100700
40. Novoselov KS, Geim AK (2007) The rise of graphene. Nat Mater 6(3):183–191
41. Tamagnone M, Gómez-Díaz JS, Mosig JR, Perruisseau-Carrier J (2012) Analysis and design of terahertz antennas based on plasmonic resonant graphene sheets. Journal ofappliedphysics 112(11):114915

42. Löffler T, Siebert K, Czasch S, Bauer T, Roskos HG (2002) Visualization and classification in biomedical terahertz pulsed imaging. Phys Med Biol 47(21):3847

43. Rabbani MS, Ghafouri-Shiraz H (2015) Size improvement of rectangular microstrip patch antenna at MM-wave and terahertz frequencies. Microwave Opt Technology Letters, 57(11), 2589.

44. Hirata A, Kosugi T, Takahashi H, Yamaguchi R, Nakajima F, Furuta T, Ito H, Sugahara H, Sato Y, Nagatsuma T (2007) 10-Gbit/s wireless communications technology using sub-terahertz waves. In Terahertz Physics, Devices, and Systems II (Vol. 6772, p. 67720B). International Society for Optics and Photonics

45. Strinati EC, Barbarossa S, Gonzalez-jimenez JL, Kténas D, Cassiau N (2019) 6G : The next frontier 42–50. https://doi.org/10.1109/MVT.2019.2921162

46. Ghafoor S, Boujnah N, Rehmani MH, Davy A (2020) MAC protocols for terahertz communication: A comprehensive survey. IEEE Communications Surveys & Tutorials 22(4):2236–2282

47. Aqlan B, Himdi M, Vettikalladi H, Le-Coq L (2020) Sub-THz circularly polarized horn antenna using wire electrical discharge machining for 6G wireless communications. IEEE Access, 8, 117245

48. Lamminen A, Säily J, Ala-Laurinaho J, de Cos J, Ermolov V (2020) Patch antenna and antenna array on multilayer high-frequency PCB for D-band. IEEE Open Journal of Antennas and Propagation 1:396–403

49. Ye N, Han H, Zhao L, Wang AH (2018) Uplink Nonorthogonal Multiple Access Technologies Toward 5G: A Survey. Wirel Commun Mob Comput 2018. https://doi.org/10.1155/2018/618 7580

50. Stoica RA, de Abreu GT (2019) 6G: the wireless communications network for collaborative and A.I. applications. arXiv:1904.03413.

51. Hussain F, Hassan SA, Hussain R, Hossain E (2020) Machine learning for resource management in cellular and IoT networks: potentials, current solutions, and open challenges. IEEE Commun Survey Tutorials

52. Zhang Z, Wang L, Liu W, Yan Z, Zhu Y, Zhou S, Guan S (2020) in Sc in It Sc. 19104, 1–10

53 Singh P, Singh N (2020) Blockchain With IoT and A.I.: a review of agriculture and healthcare. Int J Appl Evolution Computat (IJAEC), 11(4), 13–27

54 Sekaran R, Patan R, Raveendran A, Al-Turjman F, Ramachandran M, Mostarda L (2020) Survival study on blockchain based 6G-enabled mobile edge computation for IoT automation. IEEE Access 8:143453–143463

55 Shahraki A, Abbasi M, Piran M, Taherkordi A (2021) A comprehensive survey on 6g networks: applications, core services, enabling technologies, and future challenges. arXiv:2101.12475

56 Verma A, Singh P, Singh N (2021) Study of blockchain-based 6G wireless network integration and consensus mechanism. Int J Wireless Mobile Comput 21(3):255–264

57 Li W, Su Z, Li R, Zhang K, Wang Y (2020) Blockchain-Based Data Security for Artificial Intelligence Applications in 6G Networks. IEEE Network 34(6):31–37. https://doi.org/10.1109/MNET.021.1900629

# Key Technologies and Architectures for 6G and Beyond Wireless Communication System

**Aman Kumar Mishra and Vijayakumar Ponnusamy**

**Abstract** 6G and beyond wireless networks will satisfy the requirements of a fully connected world and thereby provide 'truly ubiquitous 'connectivity everywhere. This chapter will discuss the following key technology for 6G and beyond wireless network: Massive Cell-Free MIMO, Quantum Communication, Intelligent Communication environment, Internet of Nano Things, Internet of Space Things with CubeSats, Artificial Intelligence for 6G and beyond Network, Terahertz band Communication, UAV-Based Communication, Visible Light Communication, Ambient Backscatter Communication, Context-aware cognitive radio, Reconfigurable Transceiver Front-ends for 6G and beyond, Advanced 6G Applications, Research and Project Activities. Each topic is discussed at length.

**Keywords** 6G and beyond · Cell-free massive MIMO · Quantum communication · Artificial Intelligence · Internet of Things

## 1 Introduction

In this chapter, different technologies envisioned for 6G and beyond wireless networks are discussed. We start with cell-free massive MIMO(Multiple Input Multiple Output), which solves the issues faced by the current network's cell-centric (small cell) approach by providing 'truly ubiquitous connectivity everywhere. The next-generation network is slated to involve very complex computations due to the number of users and various applications; this could be solved by quantum computing (leading to quantum communication). The current wireless network faces propagation issues in the wireless environment, thus limiting its performance. An intelligent communication environment will enable the environment to become 'smart' by

A. K. Mishra (✉) · V. Ponnusamy
Department of Electronics and Communication Engineering, SRM Institute of Science and Technology, Chengalpattu 603203, India
e-mail: aa4581@srmist.edu.in

V. Ponnusamy
e-mail: vijayakp@srmist.edu.in

© The Author(s), under exclusive license to Springer Nature Singapore Pte Ltd. 2022    25
M. Dutta Borah et al. (eds.), *AI and Blockchain Technology in 6G Wireless Network*,
Blockchain Technologies, https://doi.org/10.1007/978-981-19-2868-0_2

employing unique materials in the surrounding facilitating hassle-free propagation of electromagnetic waves. Migration to a higher level in the spectrum reduces the wavelength of EM waves which enables the sizes of devices to be reduced and is used in very advanced applications like nano cameras, among others. Due to terrain, it is not possible to provide network infrastructure everywhere; CubeSats (a small satellite) can be employed to give seamless connections to not only those terrains but also applications like monitoring events and mobility, among others, making the Internet of Space Things (IoST) pivotal for 6G and beyond the network. The current wireless network faces many complex optimization problems which can be solved very easily by employing machine learning or deep learning (as suggested by the literature) algorithms making artificial intelligence extremely crucial for 6G and beyond the network. Then we discuss different key 6G and beyond technologies such as Terahertz band Communication, UAV-Based Communication, Visible Light Communication, Ambient Backscatter Communication, Context-aware cognitive radio and Reconfigurable Transceiver Front-ends for 6G and beyond.

Advanced 6G Applications like holographic and remote health care are discussed. Finally, the chapter is concluded by discussing several 5G and beyond research initiatives by governments and institutions across the globe.

## 2   Cell-Free Massive MIMO

Massive Multiple Input Multiple Output (mMIMO) has many advantages over conventional MIMO. For instance, mMIMO has at least ten times better spectral efficiency and much higher spatial resolution than the latter. However, mMIMO suffers from large variations in signal-to-noise ratio between cell center and cell edge users, leading to poor service to cell edge users. This issue can be solved using the smaller cell, wherein a base station/Access Point (AP) serves a particular smaller geographical area. However, the installation of smaller cells leads to very high inter-cell interference, which mars the network's performance. Therefore, a paradigm shift of the network is needed from a cellular network to a cell-free network. Recently, cell-free massive MIMO has been proposed, which overcomes the issues mentioned above. In cell-free massive MIMO, all the APs are connected to the Central Processing Unit (CPU) via parallel fronthaul links and jointly serve the User Equipment (UE) in a particular geographical area without considering any cell boundary. In the next section, we discuss different architectures of cell-free massive MIMO.

### 2.1   Architectures of Cell-Free Massive MIMO

Broadly, there are four architectures of cell-free massive MIMO based on different degrees of cooperation among APs [1]. They are classified as given below.

a.   Level 4 (fully centralized network)
b.   Level 3
c.   Level 2
d.   Level 1 (fully distributed).

a.   Level 4: We assume the cell-free massive MIMO consists of L APs serving
     K UEs and each AP consists of N antennas. In this form of cell-free massive
     MIMO, all the L APs sent there received pilot signals and data signals to CPU
     over parallel fronthaul links to CPU, which performs channel estimation and
     data signal detection.
         The received signal at the CPU is given by

$$
\begin{bmatrix} y_1 \\ \dots \\ y_L \end{bmatrix} = \sum_{i=1}^{K} \begin{bmatrix} h_{i1} \\ \dots \\ h_{iL} \end{bmatrix} s_i + \begin{bmatrix} n_1 \\ \dots \\ n_L \end{bmatrix} (1) \tag{1}
$$

     where $y_1$, $y_2$,..., $y_L$ are the received signals at L Aps, respectively, $h_{il}$ is the
     channel between ith UE and lth AP. $s_i$ is the transmitted symbol of ith UE, and
     $n_l$ is the noise at lth AP.
         In compact form, it can be written as

$$
y = \sum_{i=1}^{K} h_i s_i + n \tag{2}
$$

b.   Level 3: Instead of sending the received pilot signals and data signals directly
     to the CPU, here, each AP performs a local estimate of the received signal and
     passes to the CPU the same for final decoding. CPU processes the received
     signals to perform joint detection.
c.   Level 2: In level 2 architecture, the CPU performs detection directly by taking
     the average of received local channel estimates from APs. Unlike level 3, level
     2 architecture does not require any knowledge of channel statistics.
d.   Level 1: Level 1 is basically a small cell network. Each AP performs a local
     estimate by using a local channel estimate; one AP serves each UE, and no
     information is exchanged with the CPU.

## 2.2   ASO (AP Switch-Off) Technique in Cell-Free Massive MIMO

ASO techniques in cell-free massive MIMO are classified as shown below [2].

a.   Random selection ASO (RS-ASO): APs are turned off randomly and the only
     parameter that is optimized is the maximization of energy efficiency. This

scheme gives a lower bound of performance since it randomly switches off the AP making QoS very poor.

b.  Minimum propagation losses-aware ASO (MPL-ASO): The set of active APs ($M^A$) is selected for K UEs based on the minimum propagation loss. The rest of the APs can be switched off. This scheme would require APs to be switched off/on very fast, making its practical implementation very difficult since due to mobility, the propagation losses between AP and UE changes.

c.  Optimal energy efficiency-based greedy ASO (OG-ASO): It is an iterative greedy approach. Here at the ith iteration, the approach evaluates (L + 1-i) possible configuration of (L-i) active AP resulting from switching off one of them and selects the configuration which maximizes the energy efficiency. This scheme achieves the upper bound of performance when compared to other schemes.

## 2.3  Practical Deployment Issues of Cell-Free Massive MIMO

The practical adoption of cell-free massive MIMO is facing stiff issues due to the following reasons [3]:

a.  Cost and complexity of deployment—As seen from Fig. 1, each AP has to be connected to the CPU via parallel fronthaul. This leads to a huge cost burden on service providers since large numbers of long cables would have to be installed.

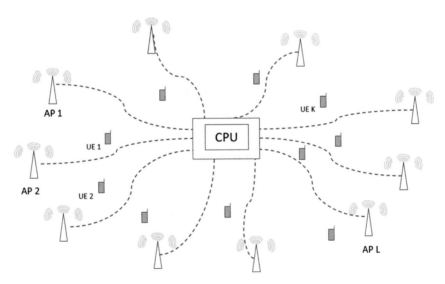

**Fig. 1** Typical Cell-Free mMIMO (Level-4)

**Fig. 2** Radio Stripe deployment scenarios

b.  Network synchronization—Achieving network synchronization would be extremely difficult in such a network since cell-free massive MIMO would involve large numbers of APs placed at different distances from the CPU.

c.  The architecture of cell-free massive MIMO suited for practical adoption is radio stripe, known as a cell-free massive MIMO system based on radio stripe. Here, APs are connected in daisy-like architecture, wherein APs are connected sequentially.

Researchers claim the adoption of radio stripe will lead to 'truly ubiquitous connectivity [3] since radio stripes can be deployed anywhere, including metro, train, public places, etc., easily given their lightweight and easy installation.

## 3 Quantum Communication

The 6G network and beyond wireless network must support millions of applications for billions of devices with far greater reliability. However, this is expected to increase the computational requirements of the wireless network manifold. Quantum computing has been recognized as a key technology to realizing the complex network. Quantum communication is defined as communication that adheres to the laws of quantum physics. Broadly speaking, quantum communication has three major advantages paving the way for its adoption in 6G and beyond the wireless network.

a.  In the era where cyber security has taken the center space owing to the risk of data hacking, quantum communication guarantees secure communication.

b.   Multiple data streams can be simultaneously encoded and transmitted.
c.   It has the potential of making very large-scale parallel computation possible.

Four rules/postulates that govern the operations of the quantum system and hence quantum communication [4] are:

Postulate 1—The quantum bit: Unlike the classical communication wherein data is represented by 0 or 1, the quantum bit or qubit is the superposition of both 0 and 1. Qubit is given by

$$|\varnothing> = a_0|0> + a_1|1> \tag{3}$$

$|\varnothing>$ is a 2D vector, $a_0$ and $a_1$ are the complex numbers, and 0 and 1 are the binary bits.

Postulate 2—The quantum Register: The quantum register is used to store qubits. However, the output of quantum registers is probabilistic; thus, the read value at the output may be different than what is stored; this imposes a challenge on the implementation of a quantum information exchange system.

Postulate 3—Exponential Speed-up: Unlike the traditional system, which employs multiple computing units for parallel processing, a single quantum computing unit can process multiple register states simultaneously. This reduces the time required for computation significantly.

Postulate 4—The Q/C conversion: It is easy to visualize the information in terms of 0 and 1, the binary bits. From (3), there is the classical interpretation that 0 would be received with probability $P_0 = |a_0|^2$ and 1 with probability $P_1 = |a_1|^2$.

## 3.1   Physical Components of a Quantum Network

a.   Quantum Nodes—These are basically the quantum devices that communicate with each other.
b.   Communication link—The communication links interconnect quantum nodes; the links may be classical or quantum links.
c.   Entanglement generator—This device is responsible for generating the entangled pairs, which are distributed among the nodes. Entanglement is the phenomenon in which the quantum state of two or more particles is described with reference to each other. Irrespective of the physical separation between particles, any action on a particle (in entangled pair) immediately affects all other particles within that pair.
d.   Quantum memories—It is used for storing quantum states for enabling communication.
e.   Quantum measurement devices—The major role of such devices is to assess the generated entangled states.

# 4 Intelligent Communication Environment

Millimeter-Wave (mmW) and Terahertz (THz) band has shown the great potential of achieving fiber-like data rate in the wireless domain and thus is slated to be adopted in 6G and beyond the wireless network. However, the communication distance is limited to a few meters owing to higher propagation losses. To solve this current issue, the state of research focuses on advanced transceiver design, completely neglecting the wireless propagation medium. The objective of an intelligent communication environment is to control how the electromagnetic waves interact with the scatters/environment. The controllable behavior of electromagnetic waves is controlled reflection, absorption, wave collimation, signal waveguiding and polarization tuning. In the subsection below, the different layers of the intelligent communication environment are discussed.

## 4.1 The Layered Structure of an Intelligent Communication Environment

Generally, an intelligent communication environment consists of four layers. These are the metamaterial plane, sensing and actuation plane, computing plane and communication plane.

a. Metamaterial plane: Metamaterial plane is basically the surface plane, as shown in Fig. 3. The metamaterial plane consists of reflect array, which has a phase shifter on every element to constructively add the useful signal and cancel interferences [6, 7]. Metamaterial tiles behave like surfaces with tunable local impedance; the incident EM wave can be routed by tuning the local impedance across the tile.

b. Sensing and actuation plane: To control the EM wave, the surface needs to sense the environment and actuate the upper surface plane accordingly.

c. Computing plane: It is an important part of the intelligent communication environment in the sense that it carries out processing functions. The

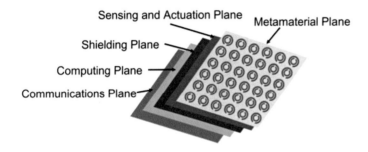

**Fig. 3** Conceptual design of a plasmonic reflectarray [5]

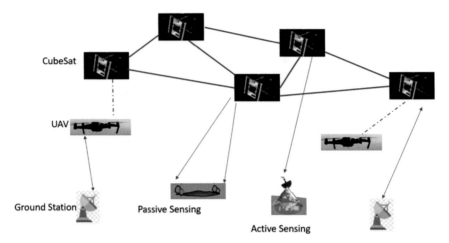

**Fig. 4** Illustration of IoST

computing plane comprises the computing hardware, which controls the actuation and sensing elements. Generally, FPGA-based controller performs these computations.

d.  Communication Plane: The communication plane primarily has two functions. Firstly, it passes the signal from the processing layer to the metasurface layer. Secondly, it collects the signal from the sensing and actuation plane. The communication plane employs command signals to communicate between planes and operate at very low frequencies. For simplicity, a communication plane is implemented within computing hardware.

## 4.2  Applications of Intelligent Communication Environment

a.  On signal propagation enhancement: There are several aspects related to EM propagation in the intelligent communication environment:

1.  Transmission distance: Intelligent communication environment can extend the coverage area in an NLOS environment. The results [8] indicate the coverage area in the NLOS area can be extended to 60GHZ.
2.  Interference mitigation: Owing to multiple users, interference becomes a cause of concern in the wireless network. Since each unit of the intelligent environment is dedicated to a particular user, the problem of interference mitigation will be no major issue in an intelligent environment.
3.  Reliability: Intelligent environments are expected to provide greater reliability and security when compared to conventional communication environments. This is due to two reasons. Firstly, the eavesdropper does not have knowledge of the frequency where packages are being transmitted.

Secondly, even if the eavesdropper is at the same frequency, it experiences much higher noise making the decoding of intercepted data impossible [9], making the dedicated link in an intelligent environment secure.

b.  On the physical layer security: The 6G and beyond wireless network is expected to provide enhanced physical layer security. Intelligent environments have been envisioned to sense the location of a legitimate user and exchange the same with the controller in order to verify the user's authenticity. Any attempt by an eavesdropper to establish a connection will be nullified. At the same time, the legitimate user will reap the benefits of an intelligent environment. An eavesdropper will receive an attenuated signal. Simulation results [10] show attenuation of 6 dB of the received signal at the eavesdropper.

# 5  Internet of Nano Things (IoNT)

Owing to the demand for a higher data rate per user and subsequent migration of internet service providers (ISP) to mmW and THz bands, the wavelength of signals falls into the nanometer range ($10^{-9}$–$10^{-7}$ m); this motivates the researchers for nano-network communication. The device employed in nano-network communication is on the scale of nanometers; thus, they are very different from traditional wireless communication devices [11]. Given such tiny sizes, each nano thing node is envisioned to be self-powered, besides being able to perform basic processing and data storage. Recent advances in nanotechnologies provide several promising materials such as thin strips of graphene named graphene nanoribbons, 3D graphene known as carbon nanotube and graphene spheres.

## 5.1  Applications of IoNT

IoNT finds applications in body area networks and short distance local environments [5]:

a.  Nano camera: Nano camera comprises nano photodetectors, nano-lens, nano-batteries and nano-memories and is used to sense, combine and process the light signal, thereby transforming them into electric signals. These cameras can be used for fracture detection in oil pipes, and intravascular imaging.
b.  On-Chip network: The wired connections on the conventional chip can be replaced by a nano-network to perform short-range communication at THz frequency.
c.  Nano-robots for IoNT: Nano-robots can be deployed to collect data that are hazardous to human beings. By forming an ad-hoc network, nanorobots can aggregate and forward data packets to gateways in IoNT.

## 6   Internet of Space Things with CubeSats

The Internet of Space Things (IoST) is a spatial expansion of the Internet of Things focusing on terrestrial uses primarily. Since the Internet of Space Things involves satellites, the performance of satellites has become very important. Traditional satellite faces severe drawbacks such as long deployment cycles, high costs and high risk exposure. CubeSats, which are a set of miniaturized satellites with size ranging from 1 to 6U ('U' denotes $10 \times 10 \times 10$ cm$^3$), has the potential of being part of the Internet of Space Things to provide ubiquitous connectivity on ground, air and space. CubeSats have several advantages over traditional satellite systems, such as lower cost, shorter development cycles, higher flexibility and scalability.

### 6.1   Application Scenarios of IoST

a.   Monitoring and reconnaissance: CubeSats has the potential to play a key role in aerial reconnaissance and monitoring. The multi-resolution camera installed on CubeSats can capture infra-red, visible and ultra-violet images, which is important for applications relating to the monitoring of terrain, disaster prevention, etc. Sensors can be mounted on the CubeSats for environmental monitoring.
b.   In-space backhaul: In remote areas where the population density is quite less, internet service providers show little interest in investing in necessary infrastructure and maintenance. IoST/CubeSats can provide connectivity in such areas since it does not require ground connections. IoST can also be employed for connectivity wherein emergency connections are needed, for example, in a place affected by earthquakes, or tsunamis damaging the existing infrastructure on the ground.
c.   Cyber-physical Integration: Future advanced applications such as autonomous driving require the sensor network on earth and space to combine their data to achieve the best performance. IoST can request data from the local sensor and combine it with its own data to ensure maximum safety and performance of autonomous driving vehicles. IoST also finds similar applications for tracking in-transit consignments. IoST can keep track of drones flying over the same area, in order to avoid collisions among them. CubeSats can combine the data from the ground controller (which has limited sight due to obstruction) and its own data.

## 7   Artificial Intelligence for 6G and Beyond Network

Researchers claim the discussion on 6G and beyond wireless networks is incomplete without applications of Machine Learning (ML)/Artificial Intelligence (AI)/Deep Learning (DL). The reason for this is very obvious. The next-generation wireless network will involve millions of diverse applications serving billions of devices ranging from D-2-D, M-2-M, IoT and handheld devices. Deep Learning (DL) has the ability to solve a problem involving complex optimization very efficiently. For example, the problem of resource allocation in heterogeneous networks, routing problems and network management are all very complex non-convex optimization problems. The literature shows these problems are solved very efficiently and easily by employing DL [12–16]. The adoption of DL is all set to bring a paradigm shift in the future wireless network. In the subsection below, important aspects relating to the application of DL in the wireless network are discussed.

## 7.1 Channel Estimation

The challenge of channel estimation in future networks (for example, Cell-Free mMIMO) would be very immense because many APs with many antennas serve large numbers of User Equipment (UE) in the uplink and downlink.

Moreover, the next-generation network is envisioned to provide uninterrupted and high-quality service (HD video live streaming) to moving UE, where channel conditions keep changing very quickly. The channel estimations in such scenarios are not only challenging but also lead to very high computational complexity. DL has been successfully applied to solve such problems in mMIMO, cell-Free massive MIMO [12–14].

## 7.2 Resource Management

Researchers have always advocated for reinforcement learning for solving the problems relating to resource allocation [5]. This can be attributed to four reasons. Firstly, the decision made by RL algorithms is highly repetitive, thereby generating a large amount of training data for RL algorithms. Secondly, the RL approach has the ability to model complex systems and decision-making policies as game-playing agents. Thirdly, for problems that are difficult to be modeled accurately, the RL frameworks can be used for such problems if there exists a reward signal that correlates with the objective. Lastly, the RL agent has the ability to optimize for any specific workload by performing well under varying conditions. Therefore, owing to the nature of problem reinforcement, learning is best suited for resource allocation/management [15, 16].

## 8  Key 6G and Beyond Technologies

In this subsection, different topics relating to 6G and beyond are discussed: Terahertz band Communication, UAV-Based Communication, Visible Light Communication (VLC), Ambient Backscatter Communication, Context-aware cognitive radio and Reconfigurable Transceiver Front-ends for 6G and beyond. These technologies will not only change the nature of communication when compared to the current world but will also ensure great user experiences in terms of ultra-low latency, very high throughput and ubiquitous connectivity.

### 8.1  Terahertz Band Communication

Owing to the dramatic rise in wireless traffic, the service providers are experiencing the 'spectrum crunch'. Moreover, in order to fulfill the requirements of very high data rates (in order of terabits per second) in the wireless domain, the service providers need to find a large amount of spectrum in the higher range of EM spectrum. Terahertz band (0.1-10THz) is envisioned as one of the key technologies to fulfill the requirements of 'extremely high data per user' in 6G and beyond networks. However, propagation losses mar its adoption for long-range communication. It is slated to be used in Terabit wireless communication, local area network and for videoconferencing among devices in a small geographical area.

The THz band can be exploited in the following scenarios:

a.  Local Area Networks: spectrum bands of 625–725 GHz and 780–910 GHz bands are suitable for short-range link communication [17]. This can also be used as a bridge to ensure the transition between fiber-optics and THz band links with zero latency.
b.  Personal area network: THz band communication can provide data rate of terabits per second; it can be exploited to communicate between devices that are meters apart. This comes in handy in indoor offices.
c.  Data Center Network: The conventional data center employs wired connections among different nodes, which results in high installation and maintenance costs. THz links can be exploited to establish seamless connectivity at ultra-high speed between nodes.
d.  Nano-Networks: Wavelength of THz band falls in the nanometer range $(10^{-9})$. Hence, THz band is best suited for nano-network given the size of devices in the nano-network being in the nanometer range. The band can be exploited in nano-networks for data exchange/transfer.
e.  Inter-satellite communication: Unlike the earth's atmosphere, the THz band is not constrained by attenuation, making the THz band very attractive for inter-satellite links. Moreover, the THz band can provide link stability even if satellites drift slightly from their orbit.

f.   Wireless network on Chip: As the trend of transceiver development gains momentum, there is a need for a higher level of miniaturization and weight reduction. THz links can replace the wired connections among different modules on the chip.

## 8.2   UAV-Based Communication

Figure 5 shows a typical UAV communication scenario where the UAV communicates with ground users (in cellular or cell-free scenario); at the backend, UAV is connected with a satellite or cloud-based network. Reliability is one of the most important performance measuring metrics in any network. There have been many incidents in the history of humankind where natural calamities such as earthquake, floods and tsunamis have caused heavy losses to communication infrastructure on the ground causing isolation from the outside world. An Unmanned Aerial Vehicle (UAV) has been envisioned to provide service to these affected areas, thus ensuring the reliability of the network. UAVs and drones are used interchangeably in the literature. UAV communication exhibits the following attractive features [18]:

a.   Line of sight links: UAVs providing service to ground users can hover at different altitudes in order to ensure that ground user experiences line of sight with it. Since, unlike base station/AP, the drone can hover in the air, it has a very high probability of having LOS links to ground users.

b.   Dynamic Deployment Ability: As per need, UAVs can be deployed to any geographical location, ensuring connectivity to ground users. Moreover, the deployment of UAVs will save the cost of service providers since they do not need to rent the sites, or install towers and cables.

c.   UAV-based swarm networks: Swarms of UAVs can form a scalable multi-UAV network to provide ubiquitous connectivity to ground users.

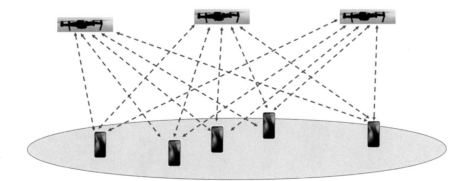

**Fig. 5**   A typical UAV/drone communication Scenario

UAVs have many applications in the modern world, like pizza and consignment delivery. The major issue that mars the adoption of UAV is their battery life. Once fully charged, UAVs can hover over different altitudes for a maximum period of 4–6 h. Moreover, the next-generation network will require the base station/AP to be installed with hundreds of antennas which seems extremely difficult for UAVs given the size and payload limitations.

### 8.3  *Visible Light Communication (VLC)*

Figure 6 shows a visible light communication scenario inside the room where a particular LED serves a particular user; due to propagation issues of light that cannot cross walls, VLC is envisioned for indoor applications and short-distance communication. Visible Light Communication (VLC) is an attractive technology for 6G and beyond wireless networks, owing to four reasons [19–21]:

a.  The band used for VLC is between 430 and 790 THz, which is an unlicensed band and hence is free to use.
b.  The band employed for VLC has a very high bandwidth.
c.  Since visible light is blocked by the walls, it has high spatial reusability.
d.  VLC supports a very high data rate (10 Gbps using LED and 100 Gbps using LASER diode) in the wireless network.
e.  The use of LEDs results in low energy consumption.
f.  It is less costly when compared to RF communication in Millimeter Wave (mmW) and TeraHertz band. This is because, unlike the former, which needs a sophisticated transceiver, VLC employs Light Emitting Diode (LED) and Light

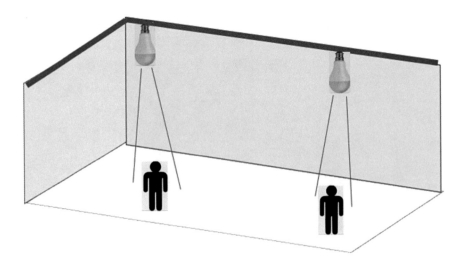

**Fig. 6**  A Visible light communication scenario inside the room

Amplification by Stimulated Emission of Radiation (LASER) diode as a transmitter, whereas it employs silicon photodiode, PIN photodiode (PD) and PIN avalanche photodiode (APD) as a detector which has comparatively very low costs.

g.   One of the most exciting features of VLC communication is it being interference-free since visible light bands are well separated.

VLC is a wireless optical technology, which attracts applications like vehicular communication, underwater communication, homes and offices, among others. VLC has matured over the last two decades for it to be considered as a technology for 6G and beyond [22] and [23]. Like THz, VLC is also best suited for short communication since visible light cannot penetrate walls.

## 8.4   Ambient Backscatter Communication

It is envisioned that next-generation wireless networks will involve large numbers of sensors for different applications such as collecting information and transmitting data to nearby devices, among others. Since these sensors are not connected to any infrastructure, they need long-lasting battery life. Radio Frequency Identification (RFID) is one such technology that can help cater to this need. RFID exploits backscatter communication techniques. In backscatter communication, the reflected signal (from RF source) has impinged on a backscatter transmitter, which modulates and sends it back to the receiver. In ambient backscatter, the communication transmitter can harvest energy from surrounding owing to radiated EM waves from base station/AP, among others, and thereby use the simple circuit for modulation for reflecting it toward the receiver [5].

However, the existing backscatter communication requires the transmitter to be placed in closed proximity among the RF sources and backscatter transmitter. It was owing to the attenuation of signals over long distances. Furthermore, in existing backscatter, the backscatter transmitter is passive in nature; that is, it cannot transmit until requested by the backscatter receiver.

## 8.5   Context-Aware Cognitive Radio

Cognitive radio was developed to solve the issues of spectrum scarcity. Cognitive radio is an intelligent radio, which aims to use the 'unused' spectrum of licensed/primary users by sensing the 'white space'. Conventional cognitive radio is envisioned to be used for a specific purpose. However, owing to the requirements of 6G and beyond wireless networks, cognitive radio should be developed for multi-scenario purposes or in other words, context-aware cognitive radio should be developed [24]. Such radio will enhance the user experience by modifying the parameters of transmission such as power, modulation and frequency channel (band). The objective of such radios is to ensure the best quality of service, reliability for end-users and is hence envisioned to be part of 6G and beyond the wireless network.

## 8.6 Reconfigurable Transceiver Front-Ends for 6G and Beyond

The next-generation radio is expected to sense and communicate over the full EM spectrum (1 GHz–10 THz), given their different applications and spectrum allocation. Limits of CMOS technology mar the prospects of reconfigurable transceivers.

### 8.6.1 Dynamic All-Spectrum Sensing and Access

'6G Mitola Radio', a research project by Research and Innovation Program in the United Kingdom, aims to establish a self-regulating society for wireless communication. This will ensure the convergence of heterogeneous networks with radio making intelligent decisions to maximize the quality of experience of end-users. However, the major challenge for this remains cross-layer control protocols and cognition framework.

One of the solutions to implementing such a transceiver is by employing software-defined radio. However, such a transceiver will lead to high energy consumption. Novel approaches should be used to use such transceiver front-ends, such as MEMS switches, nanoelectromechanical systems (NEMS), which can sense the full EM spectrum (1 GHz–10 THz). Moreover, deep learning algorithms may be employed for developing solutions for identifying free available spectrum, tuning channels and adjusting power levels [5].

## 9  Advanced 6G Applications

In this section, we discuss some very advanced applications, which can be realized in 6G and beyond networks.

a. Holographic Teleportation: Holographic Teleportation is a 3D capture technology that leverages all five senses—hearing, sight, touch, taste and smell—and allows the same 3D model to be compressed, transmitted and reconstructed anywhere across the globe. This advanced technology cannot be carted by Virtual Reality (VR) and augmented reality (AR), since holographic teleportation would require a data rate close to 5Tbps and end-to-end latency of less than 1 ms [25]. As discussed above, the THz band can provide terabits per second of data rate and is hence best suited for holographic teleportation. Holographic teleportation is envisioned for applications like real-time video conferencing at multiple locations and remote diagnosis, among others.

b. Real-time remote health care: The success of real-time health care depends on the quality and availability of connectivity [26]. It requires a very high-speed and ultra-low latency communication link, which can be fulfilled by 6G and beyond networks like THz band communication. Internet of Space Things has been envisioned to provide connectivity for rural healthcare solutions. It is envisioned that 6G and beyond networks will help doctors to treat diseases [5].

c. High-Performance Precision Agriculture: Based on the crop, soil parameters and weather, the wireless network can be exploited to take decisions regarding irrigation. For accurate time decisions, such a system will require very high data rate communication links in rural areas.

d. Intelligent Industrial Automation: Much is being talked about industrial automation, which would not only increase the industrial output but will also save costs. The upcoming Industry X.0 paradigm aims to bring Artificial Intelligence (AI) [27]. This would require high data connectivity to collect, store and process the data in real time.

e. Automation Cyber-Physical Systems: UAV/drones are one of the most promising cyber-physical systems today [5,28 and 29]. UAVs need ultra-fast communication links with high reliability for avoiding collision and communicating with infrastructure on the ground.

## 10   Research and Project Activities

There are several 6G research and development works going around the globe.

a. 6G Flagship (May 2018–April 2026): The 6G flagship project [30] is a research project funded by the Academy of Finland, which aims to develop standards for 5G and commercialization of the same, along with the development of 6G standards for the future digital society. The objective of this research is communication between devices, processes and objects to enable an automated and smart society. It plans to carry out large pilot tests with support from academia and industry.

b. HEXA-X (January 2021–June 2023): Hexa-X [31] is the European Commission flagship program for realizing beyond 5G/6G network with the vision of fully connected human, physical and digital worlds. Hexa-X is explorative research that focuses on the following:

   i. Realizing new radio access technologies at higher frequencies (mmW, THz band), sensing and high-resolution localization.

   ii. AI-driven network.

c. TERAFLOW (January 2021–June 2023): Teraflow's [32] project aims to employ software-defined networking (SDN) controller for realizing beyond 5G/6G wireless network.

d.  DAEMON (January 2021–December 2023): Daemon project [33] focuses on network intelligence (NI) for realizing beyond 5G/6G wireless network. It aims to design end-to-end NI architecture beyond the 5G network. They envision exploiting machine learning and other AI model for this purpose.

e.  6G BRAINS (January 2021–December 2023): 6G BRAINS [34] aims at exploiting deep reinforcement learning (DRL) by considering AI-driven multi-agent DRL for 6G radio. It aims to provide a cross-layer DRL-driven resource allocation framework for sub 6 GHz, THz and optical wireless communication (OWC). It aims to provide connectivity for D2D communication.

f.  South Korean MSIT 6G research program: South Korean ministry of science and ICT (MSIT) [35] is planning to launch the first 6G network in the world, expecting it to be commercially available to people between 2028 and 2030, for which it plans to invest $169 million during the period of 2021–2026, funding the necessary research and development. Five applications, namely self-driving cars, smart cities, smart factories, digital health care and immersive content, have been identified by the government for pilot projects.

## 11  Conclusions

In this chapter, we discuss different technologies and architectures slated to be 6G and beyond the wireless network. Next-generation 6G and beyond networks will go through a paradigm shift in order to satisfy the demand for end-users. Architectures such as 'radio stripe' (an implementation of cell-free massive MIMO) will provide truly ubiquitous connectivity everywhere, while quantum computation will solve many complex problems very efficiently. An intelligent environment will change the propagation characteristics of EM in the environment. The adoption of Deep Learning (DL) will solve complex problems efficiently and reduce computational complexity. Technologies like UAV can be used at the time of natural calamity to ensure the reliability of the network. To achieve terabits/second of data rate, THz band can be exploited, while Visible Light Communication can be used for achieving extremely high data rate communication for short-range communication. These technologies will change the wireless network to provide the best user experience and satisfy the expectations of end-users.

## References

1.  Björnson E, Sanguinetti L (2019) Making cell-free massive MIMO competitive with MMSE processing and centralized implementation. IEEE Trans Wireless Commun 19(1):77–90
2.  Femenias G, Lassoued N, Riera-Palou F (2020) Access point switch ON/OFF strategies for green cell-free massive MIMO networking. IEEE Access 8:21788–21803
3.  Interdonato G, Björnson E, Ngo HQ, Frenger P, Larsson EG (2019) Ubiquitous cell-free massive MIMO communications. EURASIP J Wirel Commun Netw 2019(1):1–13

4. Imre S (2014) Quantum computing and communications–Introduction and challenges. Comput Electr Eng 40(1):134–141
5. Akyildiz IF, Kak A, Nie S (2020) 6G and beyond: The future of wireless communications systems. IEEE Access 8:133995–134030
6. Wu Q, Zhang R (2018) Intelligent reflecting surface enhanced wireless network: Joint active and passive beamforming design. In 2018 IEEE Global Communications Conference (GLOBECOM) (pp. 1–6). IEEE
7. Tan X, Sun Z, Jornet JM, Pados D (2016) Increasing indoor spectrum sharing capacity using smart reflect-array. In 2016 IEEE International Conference on Communications (ICC) (pp. 1–6). IEEE
8. Liaskos C, Nie S, Tsioliaridou A, Pitsillides A, Ioannidis S, Akyildiz I (2018) A new wireless communication paradigm through software-controlled metasurfaces. IEEE Commun Mag 56(9):162–169
9. Wyner AD (1975) The wire-tap channel. Bell Syst Tech J 54(8):1355–1387
10. Liaskos C, Nie S, Tsioliaridou A, Pitsillides A, Ioannidis S, Akyildiz I (2019) There is a novel communication paradigm for high capacity and security via programmable indoor wireless environments in next-generation wireless systems. Ad Hoc Netw 87:1–16
11. Akyildiz IF, Jornet JM, Pierobon M (2011) Nanonetworks: A new frontier in communications. Commun ACM 54(11):84–89
12. Jin Y, Zhang J, Jin S, Ai B (2019) Channel estimation for cell-free mmWave massive MIMO through deep learning. IEEE Trans Veh Technol 68(10):10325–10329
13. Le Ha A, Van Chien T, Nguyen TH, Choi W (2021) Deep Learning-Aided 5G Channel Estimation. In 2021 15th International Conference on Ubiquitous Information Management and Communication (IMCOM) (pp. 1–7). IEEE
14. Bai Q, Wang J, Zhang Y, Song J (2019) Deep learning-based channel estimation algorithm over time selective fading channels. IEEE Transactions on Cognitive Communications and Networking 6(1):125–134
15. Mao H, Alizadeh M, Menache I, Kandula S (2016) Resource management with deep reinforcement learning. In: Proceedings of the 15th ACM workshop on hot topics in networks, pp 50–56
16. Zeng D, Gu L, Pan S, Cai J, Guo S (2019) Resource management at the network edge: A deep reinforcement learning approach. IEEE Network 33(3):26–33
17. Nagatsuma T, Ducournau G, Renaud CC (2016) Advances in terahertz communications accelerated by photonics. Nat Photonics 10(6):371–379
18. Shi W, Zhou H, Li J, Xu W, Zhang N, Shen X (2018) Drone assisted vehicular networks: Architecture, challenges and opportunities. IEEE Network 32(3):130–137
19. Chowdhury MZ, Shahjalal M, Hasan M, Jang YM (2019) The role of optical wireless communication technologies in 5G/6G and IoT solutions: Prospects, directions, and challenges. Appl Sci 9(20):4367
20. Karunatilaka D, Zafar F, Kalavally V, Parthiban R (2015) LED-based indoor visible light communications: State of the art. IEEE Communications Surveys & Tutorials 17(3):1649–1678
21. Jovicic A, Li J, Richardson T (2013) Visible light communication: opportunities, challenges and the path to market. IEEE Commun Mag 51(12):26–32
22. Strinati EC, Barbarossa S, Gonzalez-Jimenez JL, Ktenas D, Cassiau N, Maret L, Dehos C (2019) 6G: The next frontier: From holographic messaging to artificial intelligence using subterahertz and visible light communication. IEEE Veh Technol Mag 14(3):42–50
23. Katz M, Ahmed I (2020) Opportunities and challenges for visible light communications in 6G. In 2020 2nd 6G wireless summit (6G SUMMIT), pp 1–5. IEEE
24. Vijayakumar P, Malarvizhi S (2017) Fuzzy logic based decision system for context aware cognitive waveform generation. Wireless Pers Commun 94(4):2681–2703
25. Li, R (2018) Towards a new internet for the year 2030 and beyond. In Proc. 3rd Annu. ITU IMT-2020/5G Workshop Demo Day, pp 1–21
26. Mohapatra S, Mohanty S, Mohanty S (2019) Smart healthcare: an approach for ubiquitous healthcare management using IoT. In Big Data Analytics for Intelligent Healthcare Management (pp. 175–196). Academic Press

27. Abood D, Quilligan A, Narsalay R (2017) Industry X. 0 Combine and Conquer: Unlocking the Power of Digital. Accenture; Accenture: Dublin, Ireland
28. Bresson G, Alsayed Z, Yu L, Glaser S (2017) Simultaneous localization and mapping: a survey of current trends in autonomous driving. IEEE Transactions on Intelligent Vehicles 2(3):194–220
29. Shakeri R, Al-Garadi MA, Badawy A, Mohamed A, Khattab T, Al-Ali AK, Guizani M (2019) Design challenges of multi-UAV systems in cyber-physical applications: A comprehensive survey and future directions. IEEE Communications Surveys & Tutorials 21(4):3340–3385
30. 6G Flagship, Univ. Oulu, Oulu, Finland, 2020. Accessed: Mar. 29, 2021. https://www.oulu.fi/6gflagship/
31. Hexa-X. Accessed: Mar. 29, 2021. https://hexax.eu/
32. TeraFlow: Secured Autonomic Traffic Management for a Tera of SDN Flows. Accessed: Mar. 29, 2021. https://teraflowh2020.eu/
33. DAEMON: Network Intelligence for Adaptive and Self-Learning Mobile Networks. Accessed: Mar. 29, 2021. https://h2020daemon.eu/
34. 6G Brains: Bring Reinforcement-Learning Into Radio Light Network for Massive Connections. Accessed Mar 29, 2021. https://6g-brains.eu/
35. South Korean Ministry of Science and ICT. Accessed: Mar 29, 2021. http://english.msip.go.kr/english/main/main.do

# Blockchain-Enabled Decentralized Network Management in 6G

**Steven A. Wright** ⓘ

**Abstract** The Internet has evolved from a fault-tolerant infrastructure to support both social networking and a semantic web for machine users. Trust in the data, and the infrastructure, has become increasingly important as cyber threats and privacy concerns rise. Communication services become increasingly delivered through virtualized, software-defined infrastructures, like overlays across multiple infrastructure providers. Increasing recognition of the need for services to be not only fault-tolerant but also censorship-resistant while delivering an increasing variety of services through a complex ecosystem of service providers drives the need for decentralized solutions like blockchains. Service providers have traditionally relied on contractual arrangements to deliver end-to-end services globally. Some of the contract terms can now be automated through smart contracts on blockchains. This is a complex distributed environment with multiple actors and resources. Blockchains are not only proposed for use at a business services level but also in the operation of the network infrastructure including dynamic spectrum management, SDN and resource management, metering and IoT services. Traditional approaches to network management have relied on client–server protocols and centralized architectures. Digital transformation at both network operators and many of their customers has led to a software-defined infrastructure for communication services, based on virtualized network functions. Decentralized approaches for network management have gained increasing attention from researchers. The operators increased need for mechanisms to assure trust in data, operations and commercial transactions while maintaining business continuity through software and equipment failures, and cyberattacks provide further motivations for blockchain-based approaches.

**Keywords** Zero Touch Service Management · Zero Trust · Blockchain · Trust · Censorship-resistant · IoT

S. A. Wright (✉)
Georgia State University, Atlanta, GA, USA
e-mail: swright22@gsu.edu

© The Author(s), under exclusive license to Springer Nature Singapore Pte Ltd. 2022
M. Dutta Borah et al. (eds.), *AI and Blockchain Technology in 6G Wireless Network*,
Blockchain Technologies, https://doi.org/10.1007/978-981-19-2868-0_3

# 1   Introduction

The Internet has evolved from a fault-tolerant infrastructure to support both social networking and a semantic web for machine users. 6G radio technologies bring network performance improvements (e.g., in bandwidth and latency) that promise not only performance improvements for existing applications but also the possibility of enabling new types of applications at scale. Haptic technologies are perhaps the most easily recognizable new category of applications because they bring a new sensory mechanism (touch) to communications. Touch is a human sensory mechanism, but machine users of the Internet have different communication needs. The scale of deployments and breadth of applications for machine uses of network technologies are larger than human communication deployments, and billions of devices are already deployed in the Internet of Things (IoT). Human communication through networks has evolved through telegraph, voice, video and now haptic services requiring significantly more data. The data generated and analysed through distributed IoT devices is also expected to grow significantly [1]. Progress in analytics, and semantic annotation of available data, also supports increased machine uses of communication services. The law of large numbers dictates that failures are inevitable in systems of this scale. Sophisticated architectures and redundant deployments have largely constrained the scope of physical failures. As communications have become more data-centric, and used by machines rather than humans, issues related to data failures (e.g., incomplete, improperly formatted, unverifiable and untrustworthy data) become more problematic.

Trust in the data, and the infrastructure, has become increasingly important as cyber threats and privacy concerns rise. Trust and reputation have historically been human evaluations, but these concepts are increasingly applied in computational and communication contexts. Trust in wireless networks has been studied for more than 10 years [2]. Trust is an increasingly important characteristic of IoT applications [3]. Artificial Intelligence (AI) approaches have become increasingly important in evaluating trust computationally [4]. Blockchains have also gained broad attention as a trust machine design pattern. Trust is critical in the uncertain and evolving relationships between entities. As applications, including network management, evolve from centralized monolithic architectures towards decentralized microservice architectures, the number of communicating entities necessarily increases and so does the importance of trust in those communications. Network management is not just the trustworthiness of the data it consumes, but also the management of trust relationships required for the proper operation of the infrastructure.

Communication services become increasingly delivered through virtualized, software-defined infrastructures, like overlays across multiple infrastructure providers. Virtualized infrastructures are based on some model or abstraction of the underlying resources. That model may have limited accuracy, precision or range of operating validity. The underlying resources are multidimensional (e.g., computing, storage and bandwidth), distributed and heterogeneous in nature. While there have been significant theoretical efforts associated with optimizing these resources, experimental approaches to characterizing the resources can also be expedient [5]. Software-defined infrastructures enable automated configuration of those infrastructures, but the optimal degree of configurability and level of resource granularity remain open issues [6]. This has not, however, precluded applications of large-scale virtualized infrastructures in 5G [7]. Traditional networks have always required multiple operators to provide a global infrastructure. 5G networks. The higher bandwidths of 5G (and 6G) come through higher frequency carriers at a trade-off of a more limited range, resulting in a bias towards deployments of small cells as infrastructure. A variety of different business models have been proposed to support large-scale deployments of small cells [8]. The net result seems likely to be an increased number of infrastructure providers that need to be dynamically assembled in order to provide global services. This adds to the complexity of the tasks in the management of such external, decentralized resources.

Increasing recognition of the need for services to be not only fault-tolerant but also censorship-resistant, while delivering an increasing variety of services through a complex ecosystem of service providers, drives the need for decentralized solutions like blockchains. Communication infrastructures have been designed and developed to be resilient in the face of a wide variety of physical and environmental disasters (e.g., hurricanes, tornados). As the global economy has become more data-driven, it has become more sensitive not just to infrastructure availability but also to the validity of the information being passed. Censorship of communications for political, commercial or personal reasons can be viewed as an attack on the integrity of the information infrastructure, but the scope and scale of such attacks is a subject of ongoing study [9]. These types of attacks seem likely to increase rather than fade away, and hence infrastructure architectural approaches that provide resilience in the face of censorship attacks seem likely to be needed in the 6G context. Blockchain infrastructures have been proposed as a mechanism to provide some degree of censorship resistance. Whether through deletion or distortion of data, censorship is a fundamental threat to communication integrity. Detecting, measuring and controlling such threats is a significant challenge for communication networks to manage.

Service providers have traditionally relied on contractual arrangements to deliver end-to-end services globally. With 6G, the variety of services and the number of service providers are both expected to increase. This leads to pressure for automated mechanisms for service providers to dynamically assemble the required service partners to deliver services to their customers. With the rise of IoT and other machine users, APIs and other automation become necessary to enable service delivery. Blockchains have been proposed as a mechanism for financial settlements between remote parties. Some of the contract terms (e.g., liquidated damages in the event of service failures) can now be automated through smart contracts on blockchains. Automation in the contracting for service invocation and modification (e.g., via APIs) may be a novelty for communication services today. As the proportion of machine users increases in 6G, however, it becomes more of a necessity. This automation of contracting activities increases the scope of network management responsibilities.

Upgrading infrastructure and evolving services while operating at a large scale is a complex task. Virtualization of infrastructure enables data-driven automation.

Without trustworthy data to drive that software automation, the management operations of the service provider can be subverted. Automation can also exacerbate the
scope and scale of failures. Carefully targeted automation can help overcome fragility
in the infrastructure. The inevitability of failures in large-scale systems emphasizes
the importance of automated mechanisms for failover and service restoral. Decentralization eliminates the single point of failure in centralized systems. With 5G
deployments commencing now, 6G deployments are targeted to start in the 2030s.

Framing this discussion on blockchain-enabled decentralized network management in 6G requires an understanding of the 6G context. The following section
provides an overview of 6G with an emphasis on the scope of the resource and service
management required. Section 3 reviews the evolution of decentralized network
management technologies and architectures. Blockchain technology was initially
developed to support fintech applications, particularly transactions between remote
parties. While network management has always had an implied concern with operational costs, the use of blockchain infrastructure in communication markets is a
disruptive approach to what have been heavily regulated markets. Economic models
for network management in 6G are considered in Sect. 4. Smart contracts have been
used to automate financial transactions. Section 5 considers the role of smart contracts
and APIs in decentralized network management. Smart contracts are but one form
of automation. The volume, variety, velocity, value and veracity of the data driving
network management in large communication infrastructures is a clearly a big data
problem requiring automated support. Artificial Intelligence (AI) is often proposed
as a mechanism to support automation in the context of big data problems. The role
of AI in decentralized network management is reviewed in Sect. 6. While blockchain
approaches may be promising, that does not mean they are without challenges in their
design, implementation and operation. Section 7 looks at the challenges for network
operators in decentralizing network management with blockchains. Conclusions are
then presented in Sect. 8.

## 2 Scope of Network Management in 6G: Service Evolution and Resource Versus Service Management

The scope of network management in 6G is driven by expectations of the context
in terms of infrastructure and the commercial environment. 5G infrastructures
have already evolved into a complex distributed environment with multiple actors.
Communication infrastructure has evolved from service-specific infrastructures (e.g.,
telegraphy, telephony) to a more complex mix of services. The communication
services in the 6G era are expected to be more diverse for both human and machine
users.

As the complexity of the infrastructure and business environment has grown,
network management has been evolving from managing devices to more abstract
models—managing resources and services. The initial data models and information

models were very focused on device capabilities, but virtualization of the infrastructure through NFV/SDN brought increased abstraction from the underlying hardware, and software implementations executing on generic computing infrastructure. Industry forums have specified and standardized new network management architectures based on Software Defined Network, Network Function Virtualization, and orchestration or automation of many networks provisioning functions [10]. The scope of network management includes not only the physical layer (radio, fibre, etc.) infrastructure but also the virtualized network functions and the computing infrastructure they execute in. The range of infrastructure functions being automated or orchestrated is expected to increase in the context of 6G. These new paradigms have had a great impact on the legacy Network Management Systems and the Operations Support System (OSS). The increased complexity of the business environment required more automation of business support services associated with operating the commercial infrastructure (e.g., quoting, ordering and billing). Industry responded with architectural frameworks such as TMFs Open Digital Architecture (ODA) [11] and the more service-specific Lifecycle Service Orchestration (LSO) from the MEF [12]. Frameworks like these help enable end-to-end automation across multiple service providers—including those delivering services through 5G and 6G wireless interfaces. The new spectrum deployed for 5G and 6G is at higher frequencies, with a correspondingly shorter range, favouring small cells. The specialized services in 5G and the small cell focus reinforce private network deployment models rather than basic wide-area public services. The need to deliver end-to-end services through a variety of private networks reinforces the need for these architectural frameworks in the 6G context. The scope of services to be managed in the infrastructure includes not only the traditional connectivity services but also these additional automated business support functions.

The trend from universal service-to-service fragmentation is expected to continue in 6G. Virtualization of the infrastructure with NFV-SDN makes prevalent the concepts of network overlays, network underlays and network slices. Blockchain implementations of SDN controllers have been proposed for increased security [13]. The option of virtualized infrastructure also enables virtual operators that rely on under infrastructure service providers. Current NFV-SDN implementations enable network slicing, but are not adapted to support business agreements among mobile network operators (MNOs). The variety of services provided in the current 4G network already includes content services (e.g., video) and security services (e.g., Identity Management). The increasing deployment of IoT devices and edge computing can be expected to increase the requirement for service providers to offer a variety of non-traditional, trusted services. The new services in 5G and 6G (e.g., URLLC) use network slices to target industry verticals such as automobiles or smart cities. Network slices can be independently administered, adding complexity to the network management scope [14]. As these service-specific network slices reach deployment and maturity, these new forms of service diversity add further variety to the business models that may need to be supported by the service provider. Increasingly diverse, and dynamic, business models are expected, with many service providers relying on other service providers to deliver end-to-end services. In this 6G context, service management, not just resource management, becomes increasingly required for end-to-end services to be provisioned, assured and appropriate revenue collected by the parties involved. Blockchain approaches have been proposed for a variety of billing and settlement use cases [15]. Industry standards to support such blockchain use cases are also starting to emerge [16] as are open-source implementations [17]. Service fragmentation in 6G creates a more complex commercial environment requiring a corresponding increase in the scope of network management required to deliver those services.

Network management's scope has traditionally included categories such as Fault, Configuration, Accounting, Performance and Security (FCAPS). Service assurance in this context has traditionally devolved to availability, but that may no longer be adequate for 6G. Governments, enterprises and individuals are becoming increasingly dependent on data-driven information systems. Information access design patterns are also trending to embrace zero-trust network architectures [18]. Such approaches required finer grained controls over data access. Such controls also need to be managed efficiently at scale. With cybersecurity threats on the rise, the risk of data tampering, poisoning or censorship gains increasing attention as well as privacy risks associated with unauthorized or unwelcome data exposures. These can be viewed as failures of the information infrastructure. In the 6G era, it seems that service providers will need to provide network management service assurance beyond availability including aspects such as identities, trustworthiness and censorship resistance. The scope of 6G network management needs to expand to include the trustworthiness and censorship resistance of the data underlying the delivered services and the various identities engaged in service delivery. The scope of 6G network management responsibility in configuration and accounting may remain like current practices; albeit with new resources and services to be configured and accounted for. The scope of network management responsibilities in fault, performance and security would seem to be increased with attention to new dimensions for data resources and services. 6G network management will require new metrics, measurement and control techniques to provide service assurances in data dimensions such as privacy, trustworthiness and censorship resistance.

Traditional approaches to network management have relied on client–server protocols and a centralized architecture for the network manager. Blockchains are not only proposed for use at a business services level but also in the operation of the network infrastructure including dynamic spectrum management, SDN and resource management, metering and IoT services. [19] proposed a unified SDN and blockchain architecture with enhanced spectrum management features for enabling seamless user roaming capabilities between MNOs. [20] proposed a tokenized model for sharing spectrum and infrastructure using smart contracts. Centralized approaches to network management provide operational efficiencies by aggregating analysis and control for a single service provider. The infrastructure scale and service complexity expected in the 6G era argue for network management perspectives being available to the multiple parties required for service delivery. With multiple parties required to cooperate for service delivery, mechanisms are needed to support trust and consensus in operations. In this context, traditional centralized network management approaches seem likely to be displaced by more decentralized network management approaches.

Blockchains have been proposed for a variety of roles in the 6G network [21]. Blockchain-enabled decentralized network management can be viewed from the two perspectives—decentralized network management needs to administer blockchains as a resource, or the decentralized network management itself is implemented using blockchain technology. For the purpose of scoping the 6G network management tasks, however, it is more pertinent to focus on the resources being administered and the services being delivered. The following subsections explore further the scope of 6G resource management and 6G service management.

## 2.1  6G Resource Management

The infrastructure resources to be managed in 6G can be categorized in various ways, while 6G is associated with radio infrastructure for wireless services. Much of the core network infrastructure has traditionally been non-radio communication infrastructure. In addition, the introduction of NFV brings computing infrastructure resources into the purview of network management, and edge computing exacerbates this trend. Blockchain has been proposed for resource management and sharing in 6G using multiple application scenarios, including Internet of Things, device-to-device communications, network slicing and inter-domain blockchain ecosystems [22]. 6G infrastructure resources will need to include a large base of non-radio communication infrastructure, 6G radio resources and 6G computational resources.

Network management resources of non-radio communication infrastructure in the 6G era are traditional network management resources. Communication links (e.g., fibre optics) and switches or routers provide a traditional layer 2 and 3 connectivity between the cell towers, small cells and other nodes of 6G architectures. Information models for these resources are well developed in the industry. Multiple centralized protocols (e.g., SNMP) exist for operating networks based on these information models. These resources are typically managed by a single operator. Non-radio communication infrastructure is expected to remain a significant infrastructure base providing connectivity and other services to 6G radio resources at the edges.

Network management of 6G radio resources provides the opportunity for management of dedicated (licensed) spectrum as well as shared (unlicensed) spectrum. Spectrum licensing has allocated frequency bands within specified geographic areas to specific service providers. These independent frequency band selections in different geographic areas provide a basis for decentralized solutions. Unlicensed spectrum may be used by multiple parties—typically accepting the risk of interference. Even licensed spectrum may not always be used—offering the potential for innovation. To leverage such spectrum whitespace typically requires regulatory or licensing changes, and also, coordination mechanisms to reduce interference. The deployment of small cells, particularly in private networks, provides a specific context for spectrum sharing that seems likely to be prevalent in 6G. Blockchain mechanisms have been proposed for spectrum sharing. [23] proposed blockchains as a trusted mechanism for spectrum sharing between the multiple network operators. The large-scale deployment of IoT infrastructure is also driven by multiple parties in a distributed and decentralized pattern. ITU-T Y.4464 provides a framework for a Blockchain of Things as a decentralized service platform [24]. Network management of 6G radio resources seems likely to require decentralized solutions.

Network Management of computational resources in the 6G era perhaps requires a new resource type for some communication service providers. Management of cloud computing resources is, however, a well-known technology. Cloud computing is typically a centralized resource, and several service providers have developed their own cloud infrastructure or leveraged public cloud infrastructures. Edge computing, however, creates a new, and more decentralized, computing resource deployment to support data-intensive, low-latency 6G services. Physical computing resources may be centralized in data centres or distributed at the edge. The service provider may operate the computing resources directly, or outsource this to another (e.g., cloud) service provider. The service provider may deploy VNFs for its own services in those computing resources or offer them to other service providers. 6G services targeting lower latency typically require additional edge computing resources. Network management of decentralized 6G computational resources will become a necessary operational skill for service providers delivering data-intensive, low-latency 6G services.

## 2.2   6G Service Management

The range of services offered over 6G wireless that need to be managed is expected to be larger than the variety of services over existing networks. Each of those services needs to have network management support for the life cycles of the components supporting the service as well as the service instances and client operations. Beyond life cycle operations, the quality of the services delivered also needs to be assured. This requires the appropriate service-specific metrics and measurement techniques. Management of connectivity-based communication services (e.g., ethernet, Internet) is business as usual for most service providers today. These services primarily require partners to assist with the scope and scale of market coverage. More complex, data-intensive services may require additional partners to provide necessary functions. Scaling delivery may also require additional partners to provide the appropriate market coverage. Management of 6G services needs to support more complex services in a more complex commercial environment, and yet perform effectively as the services and infrastructure scale.

Industry vertical services (Retail, Manufacturing 4.0, Healthcare 4.0) provide customized network configurations optimized for specific industries. These often include additional design and assurance aspects to be compliant with industry-specific guidelines. With the advent of 5G verticals and the Internet of Things paradigm, Edge Computing has emerged as the most dominant service delivery architecture, placing augmented computing resources in the proximity of end users. The resource orchestration of edge clouds relies on the concept of network slicing, which provides logically isolated computing and network resources. The orchestration of multi-domain infrastructure or multi-administrative domain is still an open challenge. Reference [25] proposes a blockchain-based service orchestrator that leverages the automation capabilities of smart contracts to establish cross-service communication between network slices of different tenants. They also introduce a multi-tier architecture of a Blockchain-based network marketplace, and design the life cycle of the cross-service orchestration. By customizing services to industry verticals, service providers deliver greater value at the cost of having to manage a greater number of more complex services. Since these industry vertical services are not collocated, the infrastructure necessarily becomes more fragmented and decentralized. Industry vertical services imply increased scale, complexity and decentralization of the network management tasks the service provider needs to deliver these services. Figure 1 illustrates the increasing complexity of service provider relationships. In Fig. 1a, the client relationship with the service provider is the major focus, but this assumes that a single service provider can meet the client's needs. For more global services (e.g., Internet), collaboration across multiple service providers is needed and this has traditionally been achieved through manual curation of a set of commercial relationships by the service provider, who then masks these relationships from the consumer as shown in Fig. 1b. As the variety of services becomes both more fragmented (with more types of service providers) and more dynamic (e.g., transactional rather than subscription), the commercial trust relationships become more complex

(c) Client Service Provider Relationship    (b) Manually Curated Service Provider Relationships

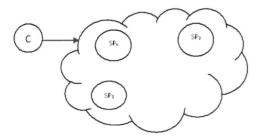

(a) Dynamic Service Provider Relationships

**Fig. 1** Service provider Relationship models

and dynamic. Emerging crowdsourced infrastructure (e.g., Helium network, https://www.helium.com) provides an early example of the increasing variety of service providers (as shown in Fig. 1c) that consumers of services will need to interact with.

The management of trusted services (identity services, trusted computing, privacy services) requires both an understanding of the appropriate trust models and also the threats to those models. These services are primarily concerned with access to data and other resources. These services also often rely on other service partners for specific functionality, e.g., specific data sets and tools. With the expected dominance of IoT devices in 6G wireless networks, the trustworthiness of the IoT data becomes a significant challenge. There has been little work on trust computation in IoT environments for service management. The case of misbehaving owners of IoT devices that provide services to other IoT devices in the system seems to be one of particular concern. Reference [26] classified existing trust computation models for service management in IoT systems based on five essential design dimensions for a trust computation model: trust composition, trust propagation, trust aggregation, trust update and trust formation. To cope with cybersecurity threats, most 6G services will require delivery in a zero-trust context. The diverse, and dynamic nature of the commercial relationships at the invocation of most 6G services, requires the infrastructure to provide automated protocols to establish and maintain the trust underlying these commercial relationships.

Management of 6G services, uniquely enabled by 6G (e.g., haptics), often requires specific I/O devices for the human interface to these services. The new human I/O devices for these services will require management to assure proper operation. Many 5G, and 6G, services are also provided for machines—e.g., IoT devices. The service assurance task for machine users is qualitatively different than for human users. As a trivial example, machines typically do not call the service providers' help desk and ask for assistance. While service providers seek to minimize problems that could result in human call help desks, the identification and resolution of troubles, at scale, for machine users is a significantly harder task. The new services uniquely enabled by 6G will require management. This may be required to be zero-touch management to enable operational scale and consumer ease of use; but increased administrative functionality will also be required for zero-trust environments.

## 3   Evolution of Decentralized Network Management Technologies and Architectures

Decentralized approaches for network management have gained increasing attention from researchers. In 1997, Baldi et al. [27] recognized the rigidity of centralized client–server architectures for network management, but they approached the problem of decentralized network management by emphasizing code mobility as a design paradigm. Mobile agents performing network management distribute some of the data collection bandwidth and data processing but do not provide a consolidated view of the network resources and services being managed. In 2009, Brunner et al. [28] proposed a probabilistic paradigm for decentralized network management. This probabilistic approach was targeted at reducing network resources associated with not only the collection and analysis of network data, but even the authors were somewhat sceptical of the acceptability of such an approach for network operators who are used to more deterministic network management systems as well as risk-averse due to regulatory penalties from infrastructure outages.

Policy-based management has been deployed in many networks to provide a higher level abstraction of network resources. While policy-based management is not new [29], recent trends in policy-based management have been towards capturing and structuring policies to capture the operational intent. There is little work which empirically analyses network management practice with respect to policy creation and maintenance. Reference [30] carried out semi-structured interviews of network administrators to identify dimensions for network policy including User, Device, Locus, Traffic features, Physical location, Temporality, Authentication and Trigger Action. Much of the recent policy- based network management work has been targeting intent with domain-specific, declarative languages to capture the policies. This trend was intensified by digital transformation and virtualization efforts at both network operators and many of their customers, starting around 2012. This digital transformation has led to a software-defined infrastructure for communication services, based on virtualized network functions. Valocchi et al. [31] proposed recognizing the increased complexity of centralized network management in the context of virtualized network functions and software-defined network infrastructures. They proposed a signalling framework to support decentralized management and control of virtualized communication infrastructure. While this approach partitioned the workload amongst local controllers, it lacks the consensus mechanisms of blockchains for coherent views of resources—especially shared resources. Policy-based approaches for network management remain an important strategy for developing the necessary abstractions to manage the scale, and diversity of modern communication infrastructures.

The explosive growth in wireless applications has provided an impetus for new network architectures beyond fixed wireless. New network architectures such as peer-to-peer networking, mobile ad hoc networking, delay-tolerant networking and pervasive computing networks. Many of these network architectures are distributed and self-organized leading some to refer to them as autonomous networks [32]. Autonomy in networks is often conflated with the use of artificial intelligence, particularly run-time learning behaviours [33]. While significant strides in artificial intelligence have been made in recent years, cognitive autonomous networks have several areas requiring further research [34]. Measuring the various dimensions of autonomy is not a simple task as these dimensions are often derived from social or psychological studies of human behaviour rather than the physics underlying most network engineering. Some blockchain structures, Decentralized Autonomous Organizations (DAOs), are designed to operate autonomously [35]. Autonomous networks are emerging from a research topic to industrialization among commercial service providers [36]. The complexity in operating and managing 5G and beyond networks has reinforced a trend towards closed-loop automation of network and service management operations. ETSI Zero-touch network and Service Management (ZSM) framework is envisaged as a next-generation management system that aims to have all operational processes and tasks executed automatically [37]. AI-based approaches are seen as likely implementation technologies to realize these architectures. Such ZSM architectures are already being proposed in wireless networks [38]. Zero-trust networks evolve the cybersecurity paradigms to move defences from static, network-based perimeters to focus on users, assets and resources. Zero trust assumes there is no implicit trust granted to assets or user accounts based solely on their physical or network location (i.e., local area networks versus the Internet) or based on asset ownership (enterprise or personally owned). Authentication and authorization (both subject and device) are discrete functions performed before establishing a session with a resource. There are already proposals for zero-trust architectures for 5G healthcare applications [39]. These architectural trends towards *autonomy, zero touch* and *zero trust* are expected to continue as a response to networking requirements. Blockchain infrastructures seem to provide an approach to address some of these requirements.

Maksymyuk et al. [40] proposed the use of blockchain mechanisms for intelligent network management in the context of sharing unlicensed 5G spectrum and infrastructure. This approach relied on a novel blockchain token corresponding to 180 kHz of the unlicensed spectrum resource being managed. This approach did not address the breadth of network management challenges in 5G and 6G, nor does it account for other users of unlicensed spectrum that may be non-cooperative. In a broader survey on the convergence of Blockchain and 5G, Nguyen, Pathirana, Ding and Seneviratne [41] identified roles for blockchains in 5G enabling technologies (e.g., NFV-SDN), 5G services and 5G IoT applications. Management roles identified for blockchains included spectrum management, interference management and resource management. [42] surveyed a broad range of ideas related to Blockchain integration in 5G and beyond networks that address issues such as interoperability, security, mobility, resource allocation, resource sharing and management, energy efficiency and other desirable features and proposed a taxonomy of blockchain for 5G categorizing applications as network management, computing management, communication management, security and privacy, applications and services. This taxonomy reflects a very narrow scope of network management (only SDN, NFV and network slicing), while many of the other categories identified are also involved in the operation, administration maintenance and provisioning of the infrastructure to meet the business objectives of the operator. When comparing different architectural approaches for network management, the availability of suitable metrics is helpful for comparing the effectiveness of divergent solutions. Wang Lu et al. [43] provide an example of such an evaluation comparing operational quality in the context of specific centralized and decentralized network management architectures. Unfortunately, the decentralization approach they chose was based on a traditional geographic hierarchy rather than a blockchain. The operators increased need for mechanisms to assure trust in data, operations and commercial transactions while maintaining business continuity through software and equipment failures and cyberattacks provide further motivations for blockchain-based approaches.

While there are proposals for blockchain use cases, proper design techniques are required for successful applications of the technology. The application functionality needs to be allocated between the blockchain system and other components. Crowcroft [44] suggests considering the performance requirements (e.g., transaction throughput, latency) before selecting blockchains as other architectures have higher transaction performance. Large-scale applications using blockchains are likely to move significant computation and storage tasks off-chain [45]. Others [46–48] have proposed more rigorous methodologies for selecting the appropriate blockchains aligned to application characteristics and requirements. After understanding the application requirements and characteristics, an evaluation can then be made of the degree of support provided by various alternative blockchains infrastructures in areas such as permissioned versus permissionless architecture, consensus mechanism selection, performance objectives and scaling considerations. Blockchains operate using distributed ledgers to record transactions of tokens. The design considerations need to include the token life cycles and the binding mechanism between those tokens and underlying assets. In the context of decentralized network management, either those tokens are associated with the network asset being managed or the network management task becomes the administration of the blockchain itself. The use of design patterns is a common practice in engineering. Design patterns for network architectures are more commonly discussed than network management design patterns. [49] discussed APIs for NMIs and the need for application frameworks, describe and illustrate the pattern system underlying the Layla framework, detail three of its key patterns, and put the pattern system into perspective. Design patterns for applications based on blockchains are also starting to emerge [50–52]. In the case of blockchains deployed for trust assurance, design patterns for business and trust models are also needed. Trust patterns for collaborative business processes are also starting to emerge [53]. While blockchains have demonstrated considerable potential, further maturity is required for commercial practice in the network context. Given the plethora of proposals, it seems reasonable to expect maturity of the technology and design practices by the time of 6G deployments.

Deployment of decentralized network management solutions using blockchains is a commercial rather than purely technical decision. Commercial evaluations for deployment consider not only technical design choices but also an evaluation of the company's ability to deliver, operate and maintain the solution. Readiness models [54] or maturity models [55] have been used to measure organizational readiness for transformative technology deployments. Blockchain network management systems would certainly be a transformative technology deployment, but I am not aware of any widely accepted maturity models for organizational readiness to support blockchain deployments. Blockchain software engineering is an emerging area of technical expertise where an industry consensus on the required technical skills, education and experience has not yet developed [56]. From a commercial perspective, the identification and staffing of the appropriate technical capabilities is an important organizational prerequisite for the organizational maturity needed to successfully deploy blockchain solutions. Many blockchain solutions rely on open-source blockchain software. In such cases, the maturity of the community supporting the open-source blockchain also needs to be considered. Commercial judgements of organizational readiness and maturity are critical for successfully deploying network blockchains.

## 4   Economic Models for Network Management in 6G

Economic modelling implies the existence of markets with transactions with buyers and sellers of, in this case, network management services. Most blockchains are designed to support transactions recorded in distributed ledgers. One approach to categorizing economic markets for network management in 6G is to look at the structure of the market in comparison with the organization boundary of the service provider. Here, we can distinguish between markets that are internal to a single service provider; markets where a single service provider offers its services; and markets for communication resources where multiple service providers operate as buyers and sellers.

Internal markets can exist for Network Management services within service providers. For regulatory, tax and other business reasons, service providers may choose to organize themselves into multiple internal organizations that provide services to each other. For example, a corporate holding company may have multiple legal entities operating in different jurisdictions; or form different organizations to market specific services. Internal markets are often centralized rather than decentralized. Because it is all one firm, there is less need for protocols to build trust.

A recent trend has emerged for operators' communication APIs, products and services to be made available to external users through digital marketplaces. While this provides an alternative channel for human users, these digital marketplaces become essential for machine users. Many of these network service offerings today are targeted at larger enterprises or government rather than consumer markets. Consider, for example, a connected vehicle that moves between the geographic coverage areas of different service providers. A more futuristic scenario might be a DAO that relies on an underlying communication infrastructure, periodically rebidding for communication services to optimize its ongoing business costs. In the context of such digital marketplaces, multiple actors are buying or selling network services creating a need for mechanisms to support the trustworthiness of those transactions. Settlements between service providers are already a commonplace activity from roaming voice services. Industry bodies have been developing standards [16] and open-source implementations [17] of blockchains to support these settlements. This provides some operational efficiency by eliminating external clearing houses for settlement activities. Such settlement arrangements can also be considered a type of digital marketplace.

There are a variety of different types of markets. Markets for shared resources are typically separated by the type of resource. Computing resources both in centralized clouds and in edge computing can be allocated using markets. Not all economic models provide the same benefits for users in utilizing the resources; nor in the profit earned by resource providers. Economic models can be effective in the collaborative use of large-scale heterogeneous resources that are typically owned by different organizations. Economic models may differ from one another as they are used for interaction among users and providers; they are used for pricing, and they adapt to evaluate different requirements. [57] studied diverse market models for allocating resources in the context of grid computing including commodity market, double auction, English auction, bargaining, proportional share, proportional resource sharing, first price sealed bid and contract net protocol. Markets for networking resources or services can use any of these approaches, subject, of course, to regulatory constraints.

Blockchain technology can reduce transaction costs, generate distributed trust and empower decentralized platforms. Those decentralized platforms may provide a foundation for novel decentralized business models. In the financial industry, decentralized financial services tend to be more decentralized, innovative, interoperable, borderless and transparent. [58] identified decentralized financial business models: decentralized currencies, decentralized payment services, decentralized fundraising and decentralized contracting. Not all network management functions have a financial aspect, but the service fragmentation expected in the 6G era argues for increased use of markets for different services and resources. The economics and operation of blockchain systems can significantly differ between permissioned (consortium) and permissionless (public) blockchain infrastructures. In the networking context, different markets may require different blockchain features. Markets for settlements amongst service providers may have relatively few participants and favour a permissioned consortium blockchain. Marketplaces for consumer network services based on blockchains may be better suited to public, permissionless blockchains.

Blockchains perform transactions using tokens. Decentralized finance applications rely on tokens as a form of financial asset (e.g., a crypto currency, security or commodity). Tokens as financial assets imply a corresponding regulatory treatment as financial assets. Settlement applications in the telecom industry may use existing financial cryptocurrencies, but could also be designed to use some other more industry-specific usage measures (e.g., call data records of usage minutes, and bandwidth). Tokens for other more novel applications (e.g., spectrum sharing) could also be created. The token life cycle operations (e.g., creation, deletion, partition and transfer) can significantly affect the economic significance of the token. The regulatory classification of blockchain tokens is still emerging. Inadvertent regulatory classification of tokens could have significant unanticipated commercial consequences (e.g., taxes or reporting burdens).

# 5   Role of Smart Contracts and APIs in Decentralized Network Management

Smart contracts are programs executing in a decentralized blockchain which permits them to leverage the underlying blockchain's consensus mechanism to provide computation results with a degree of resilience to failures, tampering and censorship attempts. While several blockchains support smart contracts, there is a cost to doing so in the internal bandwidth, block storage and compute resources required to support these smart contracts. In many cases, off-chain computations or storage may be used for efficiency, with the blockchain primarily used to store signatures of the results of the computations. Note that off-chain computations may have additional requirements (e.g., for confidential computing technologies) in order to preserve end-to-end service assurance objectives. Smart contracts were originally proposed in the context of decentralized finance applications where they could automate the execution of portions of financial contracts. Smart contracts are not self-activating; they only act in response to a transaction on the blockchain. Oracles are required to bring off-chain events into the blockchain context. In the network management context, the potential volume of events to be monitored and off-chain devices to be controlled is significantly greater than the data series typically used for the standardized financial contracts (e.g., options contracts). Smart contracts operate on digital assets, tokens that are supported by the underlying blockchain. Those tokens can be associated with physical assets or other digital assets. The literature increasingly shows examples of blockchain smart contracts being used for network applications.

Reviews of the smart contract literature reflect the original focus on decentralized finance applications. Smart contract applications for IoT have emerged as an important research category with network management implications. Smart contracts are usually specific to the blockchain architecture in which they are expected to execute. Ethereum is the most frequently cited blockchain within the smart contract literature, but there are other blockchains (e.g., IOTA) which are more focused on IoT applications [59]. Smart contracts have been proposed for a variety of networking-related tasks. These include spectrum sensing [60], DDoS Mitigation [61], distributed cloud storage [62], VANETS [63], software-defined networks [64] and resource allocation [65]. These network applications for blockchain smart contracts are increasingly relying on nonfinancial tokens, i.e., tokens associated with a network resource rather than a financial asset.

Smart contracts were developed to provide automated execution of some financial terms in the context of a financial contract, but network operators are no strangers to automation. As a computational mechanism, smart contracts are significantly less efficient because of the decentralization and consensus mechanisms. That efficiency is traded off for the trust assertions available through the consensus process. A common design pattern from the financial industry is to use smart contracts in decentralized settlements rather than a centralized clearing house. Such settlement arrangements certainly exist today between service providers. Other decentralized applications and design patterns requiring a strong basis of trust are also starting to emerge in the network context. The increasing diversity of service partners, and the trend towards service invocation through APIs, argue for more automated mechanisms to assure trust in transactions. The trend towards zero-trust architectures also argues for decentralized identity and authorization services than can be invoked with smart contracts. Applying zero-trust architecture in the networks of the scale of IoT further drives towards automated mechanisms to build trust relationships. The need for an automated, large-scale mechanism to support trusted networking transactions between independent entities creates a basis for expecting the wider deployment of smart contracts.

# 6 Role of AI in Decentralized Network Management

Applying Artificial Intelligence to the problem of network management is not a new approach [66, 67]. Progress in computing infrastructure, data science and artificial intelligence techniques have become sufficiently mainstream for network management applications that standard bodies have started to respond with architectures and more detailed specifications. Such specifications provide a basis for further industrialization and deployment in operator networks. It is noteworthy that the range of network management tasks is also expanding from fault identification and classification to closed-loop controls where the AI system is optimizing resource usage in response to some set of policy directives [68]. A fully operative and efficient 5G network cannot be complete without the inclusion of artificial intelligence software. Existing 4G network operations are based on a reactive concept, leading to the poor efficiency of the spectrum. AI and its subcategories like machine learning and deep learning have been evolving to the point that this mechanism allows 5G wireless networks to be predictive and proactive [69]. 6G networks can be expected to build on the AI deployments from earlier deployments, and see AI deployed in a broader range of network management tasks.

6G networks not only transform and transport massive amounts of customer data, they also have the potential to generate a massive amount of data for network management purposes. AI technologies and particularly ML have the potential to efficiently solve unstructured and seemingly intractable problems by involving large amounts of data. AI role is well recognized in learning systems as a mechanism for optimization. The application of AI in wireless is moving from large scale, centralized optimizations with average overall performance, to more independent localized optimizations with the better overall performance for the end user. Network management is evolving from model-based to data-driven optimization of the network using online software-based network orchestration, NFV-SDN and proactive wireless networking. [70] surveyed AI applications on different aspects of wireless network design and optimization, including channel measurements, modelling, estimation, physical layer research and network management and optimization. The types of applications for AI deployments are changing from classification to run-time operations. The massive data transported and generated by 6G networks creates the opportunity for creative AI designs impacting network management.

Artificial Intelligence and Machine Learning have recently developed the following from academia and industry in solving very complex problems in several fields. [71] surveyed the main application areas of AI/ML in SDN and NFV-based networks. AI/ML is expected to be a pivotal technology in the deployment of self-configured, self-adaptive and self-managed networks. Challenges in realizing this potential include computational complexity and latency, computational resource requirements, access to data resources and archival storage of network management data. Multiple industry bodies are working on AI/ML for networks including 3GPP, 5G PPP, FuTURE, ITU and TIP. Standards for operations of AI systems are starting to emerge, e.g., [72], but these have not yet been tempered with a lot of diverse operational experience. AI Operations is emerging as a field of expertise as the technology moves into large-scale deployment among service providers.

## 7 Challenges for Network Operators in Decentralizing Network Management with Blockchains

Historical challenges for service providers have centred around deploying connectivity and bandwidth. Fibre optic technologies improved backbone bandwidth, while increased cellular coverage brought connectivity to greater populations. Network operators have skilfully managed multiple technology transitions in their infrastructure—1G (analog)- > 2G (digital), 2G (circuit)- > 3G (packet) and 3G (voice)- > 4G (video). 5G marked a more radical shift by catering to diverse requirements of several use cases. These requirements include Ultra-Reliable Low Latency Communications, Massive Machine-Type Communications and Enhanced Mobile Broadband. 4G - > 5G transitions the infrastructure from physical - > virtualized and the predominant usage from humans with cell phones - > IoT devices. Several technologies, such as software-defined networking, network function virtualization, machine learning and cloud computing, are being integrated into the 5G networks to fulfil these diverse requirements. These technologies, however, give rise to several challenges associated with decentralization, transparency, interoperability, privacy and security. To address these issues, Blockchain has emerged as a potential solution due to its distributed architecture capabilities including transparency, data encryption, auditability and immutability. The 5G- > 6G transition promises additional challenges from the diversity of services to be supported both for human and machine users.

The operation of communication infrastructure is business with significant scale effects. The value of the network scales increases with size, as do the cost and complexity of managing it. Evolution at scale is still an issue in the 6G era as operators seek to deliver value through an increasingly dynamic and diverse set of services while managing the costs of the underlying infrastructure. The decentralization of functionality to the edge (e.g., edge computing) exacerbates this challenge by increasing the number of locations where operational tasks need to be performed. While virtualization of many network functions enables more automated network management, there are still significant scale impacts. Consider the scale of the problem associated with managing software updates for VNFs. A typical VNF may have a development life cycle supporting updates every 6 months, with perhaps some emergency patches at more frequent intervals if security issues are found. For an edge node supporting a marketplace of 100 different types of VNFs to support the different services offered, it implies the average interval between software update operations on VNF types is $< 2$ days. If each VNF type has 100 operating instances in the edge computing node, then the average interval between updating instances is $< 1/2$ h. For a national scale operator with say 10,000 edge nodes located by cell towers, the scale of automation required to maintain the infrastructure at the latest release becomes obvious. This does not consider the additional administrative burden of moves, adds and changes, or new service deployments. The required level of automation is increasingly driven by AI. Managing AI operations at scale is a new competency that operators will need to develop. The performance of various blockchain architectures at scale is an ongoing area of blockchain research. The challenge for network operators comes from the variety of infrastructures—e.g., a blockchain for settlements may not use the same underlying technology as a blockchain for IoT devices. Operators may need to manage multiple different decentralized infrastructures to support different network and business functions.

Service requirement diversity creates a challenge for service providers. Should they attempt to "cover the waterfront", i.e., maintain broad coverage of different services as a "one stop shop"; or specialize in one or more niches, and risk them being commoditized or overcome by changes in demand? The network management challenge increases as the number of services to be managed increases. Many communication services require extensive geographic coverage. Globalization trends and the recent pandemic have emphasized the importance of remote working capabilities, often across multiple time zones. No single service provider has the resources to meet the required geographic scope, and so inter-provider commercial relationships are required to provide the required functionality. Service requirement diversity increased the challenge of managing those commercial relationships—e.g., not all service providers will be able to provide all services as they make commercial decisions about which set of services to provide based on their resources, regulatory environment, competitive pressures, etc. The challenges of service requirement diversity add complexity to the network management task as these services become more decentralized.

Business models drive deployments of infrastructure and services and will do so again in 6G. Business model diversity is already increasing in existing networks. Consider the rise of service providers that do not provide their own infrastructure, e.g., virtual mobile network operators. 5G radio characteristics and service specializations favour densification and localization deployment strategies. Private landlords may have better commercial visibility to drive 5G deployments in specific venues, e.g., entertainment, manufacturing and health campuses. Sophisticated services like AR/VR and smart environments are by their nature very localized, and reinforce the need for edge computing to reduce service latency below perceptible limits. Latency is not the only performance metric of interest, but it does provide a challenge for many general-purpose blockchain systems. Spectrum availability and service diversification in the 6G context may exacerbate these trends resulting in an increasing diversity of 6G deployment locations and business models. Service providers then have the challenge of stitching together an end-to-end service in such an environment. The network management challenges are increased by this diversity of deployment models for 6G infrastructure.

Consumer expectations for communication services are evolving from service availability towards service trustworthiness. The dependence of modern society on access to Internet data and computer systems for individuals, enterprises and governments has highlighted a variety of different threats when those systems and data are compromised. Data on individuals may be exposed for commercial gain or public embarrassment in a privacy violation. Trust in the integrity of various societal processes can be undermined by tampering with data. Censorship of data sources can also provide an implicit bias by selectively removing data access. Models and metrics for trust, privacy, censorship resistance, etc. are still evolving. Many of these are based on specific technologies rather than broader legal notions of privacy, trust, censorship, etc. Without appropriate metrics and measurement techniques, it would be difficult to design and operate the communication infrastructure to provide assurances on these points. Network management and control operations will need further progress in these areas to support 6G infrastructures and services to support evolving consumer expectations for data trustworthiness.

IoT traffic has taken on increasing importance with billions of devices deployed and new network services designed for them being deployed as part of 5G. A further expansion of IoT traffic sources and services is expected for additional types of machine users of communication services in 6G. Blockchains have been proposed as a mechanism to support the integrity of IoT infrastructures, while the scale of IoT infrastructure remains a network management challenge. Adequately designing and operating such blockchain systems requires an understanding of the threat models these blockchain systems are required to protect against. [73] provided a detailed analysis of potential security attacks and presents existing blockchain solutions that can be deployed as countermeasures to such attacks, including blockchain security enhancement solutions. Identified blockchain security issues included transaction malleability, network security, privacy, redundancy, regulatory compliance, criminal activity and vulnerabilities in smart contracts. The scale of IoT infrastructures is a challenge for network management, driving the need for zero-touch network management capabilities. The network security challenges of such large-scale distributed IoT infrastructures drive towards zero-trust mechanisms to identify compromised devices and constrain the scope of further spread.

The network security capabilities of today's infrastructure have been developed using cryptography within the context of contemporary von Neuman computer architectures. 6G era, however, is expected to start in the 2030s. By then, quantum computing technologies threaten the integrity of many current cryptographic algorithms. Quantum supremacy refers to the goal of using a quantum computer to solve a problem that no classical computer can solve in a reasonable amount of time. Several research teams have already claimed to have achieved examples of quantum supremacy. While current quantum computing capabilities may be limited, considerable progress in quantum computing technologies may be expected by 2030. Network management, including blockchains, that relies on cryptography will need to transition to post-quantum cryptographic methods. This may imply additional costs for early movers than then need to update cryptographic software. This problem of needing post-quantum cryptographic methods, however, is not unique to blockchains. Network security is expected to remain a challenging part of network management, even in a decentralized context.

Blockchains provide an infrastructure for smart contracts executing business logic code in a decentralized architecture. In this context, the execution outcomes have a basis for trust, agreed upon by the consensus of the executing nodes. Smart contracts can be viewed as a form of distributed computing; however, the software engineering body of knowledge to support professional practices in this area is still emerging. The design patterns and other practices for blockchain software engineering are expected to develop as blockchains are deployed, and operated, in a wide variety of environments. In multiparty environments, the standardization of data semantics and operational governance of systems are important matters to resolve. The communication industry is well used to the development of industry standardization to support such initiatives. Blockchains use oracles to adapt off-chain, external data sources and controls to their distributed architecture. Independent oracles as trusted data sources may be a new type of business partner required for some services. [74] proposed a taxonomy for blockchain oracles classifying them based on the data source, trust model, design pattern and interaction model. The commercial roles and responsibilities may prove more challenging as decentralized architectures often imply significant reengineering of business processes. Coordinating such process reengineering across multiple parties during the migration of existing operating services to smart contracts on blockchain infrastructures could prove challenging for many operators.

# 8   Conclusions

Blockchain-enabled decentralized network management is a disruptive change to existing network management processes. The scope and scale of the 6G network management challenge support the need for these types of network management architectures. The network context of the 6G era of the 2030s and beyond is expected to be more complex than current communication networks from both technology and commercial perspectives. As machine usage dominates over human usage, the services offered and their invocation methods become more data-centric. Even for human users, cybersecurity threats create a demand for service assurances that data is trustworthy, private and resistant to censorship attacks. Service fragmentation, resource variety and the resulting complexity in the 6G commercial environment reinforce the need for automated protocols to support trustworthy transactions at a massive network scale. AI techniques have shown promise in managing the scale of the network management data collection and analysis problem. Blockchains have also shown promise as an enabling technology for trustworthy transactions and censorship-resistant data. Both technical and commercial or organizational challenges remain before the wider adoption of these technologies. Blockchain-enabled decentralized network management provides a promising framework for considering the operational and administrative challenges expected in the 6G communication infrastructure.

# References

1. Siow E, Tiropanis T, Hall W (2018) Analytics for the internet of things: a survey. ACM Comput Surv 51(4):1–36
2. Yu H, Shen Z, Miao C, Leung C, Niyato D (2010) A survey of trust and reputation management systems in wireless communications. Proc. IEEE. 98(10):1755–1772
3. Din IU, Guizani M, Kim BS, Hassan S, Khan MK (2018) Trust management techniques for the Internet of Things: A survey. IEEE Access 7:29763–29787
4. Wang J, Jing X, Yan Z, Fu Y, Pedrycz W, Yang LT (2020) A survey on trust evaluation based on machine learning. ACM Comput Surv 53(5):1–36
5. Zheng W, Bianchini R, Janakiraman GJ, Santos JR, Turner Y (2009) JustRunIt: Experiment-based management of virtualized data centers. Proc. USENIX Annual technical conf. (pp. 18–18).
6. Roozbeh A, Soares J, Maguire GQ, Wuhib F, Padala C, Mahloo M, Turull D, Yadhav V, Kostić D (2018) Software-defined "hardware" infrastructures: A survey on enabling technologies and open research directions. IEEE Communications Surveys & Tutorials. 20(3):2454–2485
7. Bernardos CJ, Gerö BP, Di Girolamo M, Kern A, Martini B, Vaishnavi I (2016) 5GEx: realising a Europe-wide multi-domain framework for software-defined infrastructures. Trans. on Emerging Telecommunications Technologies. 27(9):1271–1280
8. Ahokangas P, Moqaddamerad S, Matinmikko M, Abouzeid A, Atkova I, Gomes JF, Iivari M (2016) Future micro operators business models in 5G. The Business & Management Review 7(5):143
9. Niaki AA, Cho S, Weinberg Z, Hoang NP, Razaghpanah A, Christin N, Gill P (2020) ICLab: A Global, Longitudinal Internet Censorship Measurement Platform. Symposium on Security and Privacy (pp. 135–151). IEEE.
10. Saadon G, Haddad Y, Simoni N (2019) A survey of application orchestration and OSS in next-generation network management. Computer Standards & Interfaces 62:17–31
11. TMF (2020) IG1167 TM Forum Exploratory Report ODA Functional Architecture. https://www.tmforum.org/resources/exploratory-report/ig1167-oda- functional-architecture-v5-0/
12. MEF, (2016) Lifecycle Service Orchestration (LSO) Reference Architecture and Framework, MEF55
13. Moges E, Han T (2020) DecOp: Decentralized Network Operations in Software Defined Networking using Blockchain. IEEE Conf. on Computer Communications Workshops (pp. 225–230). IEEE.
14. Suárez L, Espes D, Le Parc P, Cuppens F (2019, June) Defining a communication service management function for 5G network slices. European Conf. on Networks and Communications (pp. 144–148). IEEE.

15. Darmwal R (2017) Blockchain in Telecom Sector: An Analysis of Potential Use Cases. Telecom Business Review 10(1):68
16. MEF (2021) DLT-Based Commercial and Operational Services Framework – Billing, MEF W114 Letter Ballot
17. GSMA (2021) eBusiness Network retrieved Aug 4th 2021 from: https://www.gsma.com/ser vices/gsma-ebusiness-network/
18. Rose SW, Borchert O, Mitchell S, Connelly S (2020) Zero trust architecture. NIST SP 800–207. https://doi.org/10.6028/NIST.SP.800-207
19. Okon AA, Elgendi I, Sholiyi OS, Elmirghani JM, Jamalipour A, Munasinghe K (2020) Blockchain and SDN Architecture for Spectrum Management in Cellular Networks. IEEE Access 8:94415–94428
20. Maksymyuk T, Gazda J, Volosin M, Bugar G, Horvath D, Klymash M, Dohler M (2020) Blockchain-Empowered Framework for Decentralized Network Management in 6G. IEEE Commun Mag 58(9):86–92
21. Hewa T, Gür G, Kalla A, Ylianttila M, Bracken A, Liyanage M (2020) The role of blockchain in 6G: Challenges, opportunities and research directions. 2nd 6G Wireless Summit (pp. 1–5). IEEE.
22. Xu H, Klaine PV, Onireti O, Cao B, Imran M, Zhang L (2020) Blockchain-enabled resource management and sharing for 6G communications. Digital Communications and Networks 6(3):261–269
23. A. Okon N, Jagannath I, Elgendi JM, Elmirghani H, Jamalipour A, Munasinghe K, (2020) Blockchain-Enabled Multi-Operator Small Cell Network for Beyond 5G Systems. IEEE Network 34(5):171–177. https://doi.org/10.1109/MNET.011.1900582
24. ITU-T (2020) SeriesY: Global Information Infrastructure, Internet Protocol Aspects, Next Generation Networks, Internet of Things and Smart Cities, Internet of Things and smart cities and communities- Frameworks, architectures and protocols – Framework of blockchain of things as a decentralized service platform, Y.4464
25. Papadakis-Vlachopapadopoulos K, Dimolitsas I, Dechouniotis D, Tsiropoulou EE, Roussaki I, Papavassiliou S (2021). On Blockchain-Based Cross-Service Communication and Resource Orchestration on Edge Clouds. In Informatics (Vol. 8, No. 1, p. 13). Multidisciplinary Digital Publishing Institute.
26. Guo J, Chen R, Tsai JJ (2017) A survey of trust computation models for service management in internet of things systems. Comput Commun 97:1–4
27. Baldi M, Gai, S, Picco GP (1997) Exploiting code mobility in decentralized and flexible network management. Int'l Workshop on Mobile Agents (pp. 13–26). Springer, Berlin, Heidelberg.
28. Brunner M, Dudkowski D, Mingardi C, Nunzi G (2009, June). Probabilistic decentralized network management. IFIP/IEEE Int'l Symp. on Integrated Network Management (pp. 25–32).
29. Chadha R, Lapiotis G, Wright S (2002) Policy-based networking. *IEEE*. Network Magazine 16:8–9
30. Curtis-Black A, Galster M, Willig A (2017) High-level concepts for northbound APIs: An interview study. 27th Int'l Telecommunication Networks and Applications Conf. (pp. 1–8). IEEE
31. Valocchi D, Tuncer D, Charalambides M, Femminella M, Reali G, Pavlou G (2017) SigMA: Signaling framework for decentralized network management applications. IEEE Trans. on Network and Service Management, 14(3), 616–630.
32. Jiang, T, & Baras, JS (2006, April) Trust evaluation in anarchy: A case study on autonomous networks. Proc. INFOCOM 2006. 25th Int'l Conf. on Computer Communications (pp. 1–12). IEEE.
33. Simsek M, Zhang D, Öhmann DM, M, Fettweis, G, (2017) On the flexibility and autonomy of 5G wireless networks. IEEE Access 5:22823–22835
34. Mwanje, SS, Mannweiler C (2018, November). Towards cognitive autonomous networks in 5g. In ITU Kaleidoscope: Machine Learning for a 5G Future (pp. 1–8). IEEE.
35. Wright SA (2021) Measuring DAO Autonomy: Lessons From Other Autonomous Systems. IEEE Trans. on Technology and Society 2(1):43–53

36. Salem M, Imai P, Vajrabhaya P, Amin T (2021) A Perspective on Autonomous Networks from the World's First Fully Virtualized Mobile Network. IEEE Wirel Commun 28(2):6–8
37. ETSI, (2019) Zero-touch Network and Service Management (ZSM) Reference Architecture ETSI GS ZSM 002 v1.1.1
38. Bonati L, D'Oro S, Bertizzolo L, Demirors E, Guan Z, Basagni S, Melodia T (2020) CellOS: Zero-touch softwarized open cellular networks. Comput Netw 180:107380
39. Chen B, Qiao S, Zhao J, Liu D, Shi X, Lyu M, Zhai Y (2020). A security awareness and protection system for 5G smart healthcare based on zero-trust architecture. IEEE Internet of Things Journal.
40. Maksymyuk T, Gazda J, Han L, Jo, M (2019). Blockchain-based intelligent network management for 5G and beyond. 3rd Int'l Conf. on Advanced Information and Communications Technologies (pp. 36–39). IEEE.
41. Nguyen DC, Pathirana PN, Ding M, Seneviratne A (2020) Blockchain for 5G and beyond networks: A state of the art survey. J. of Network and Computer Applications. 166:102693
42. Tahir M, Habaebi MH, Dabbagh M, Mughees A, Ahad A (2020 Jun) Ahmed KI (2020) A review on application of blockchain in 5G and beyond networks: Taxonomy, field-trials, challenges and opportunities. IEEE Access. 17(8):115876–115904
43. Wang X, Liu X, Zhang B, Wu H, Chen S (2020) Research on Evaluation Methods of Decentralized and Centralized Operation Quality of Computer Networks. 4th Information Technology, Networking, Electronic and Automation Control Conf. (Vol. 1, pp. 300–305). IEEE.
44. Crowcroft J (2020) How to tell when a digital technology is not ready for you. Patterns 1(1):100001
45. Gudgeon L, Moreno-Sanchez P, Roos S, McCorry P, Gervais A (2019) SoK: Off The Chain Transactions. IACR Cryptol. ePrint Arch. 2019:360
46. Nanayakkara S, Rodrigo MNN, Perera S, Weerasuriya GT, Hijazi AA (2021) A methodology for selection of a Blockchain platform to develop an enterprise system. J. of Industrial Information Integration 23:100215
47. Scheid EJ, Lakic D, Rodrigues BB, Stiller B (2020) PleBeuS: a Policy-based Blockchain Selection Framework. In Network Operations and Management Symposium (pp. 1–8). IEEE.
48. Farshidi S, Jansen S, España S, Verkleij J (2020) Decision support for blockchain platform selection: Three industry case studies. IEEE Trans. on Engineering Management 67(4):1109–1128
49. Keller RK, Tessier J, von Bochmann G (1998) A pattern system for network management interfaces. Commun ACM 41(9):86–93
50. Xu X, Pautasso C, Zhu L, Lu Q, Weber I (2018) A pattern collection for blockchain-based applications. In 23rd European Conf. on Pattern Languages of Programs (pp. 1–20).
51. Gasparič M, Turkanović M, Heričko M (2020) Towards a Comprehensive Catalog of Architectural and Design Patterns for Blockchain-Based Applications-A Literature Review. In Central European Conference on Information and Intelligent Systems (pp. 259–266). Faculty of Organization and Informatics Varazdin.
52. Rajasekar V, Sondhi S, Saad S, Mohammed S (2020) Emerging Design Patterns for Blockchain Applications. In Int'l Conf. on Software Technologies (pp. 242–249).
53. Müller M, Ostern N, Rosemann M (2020) Silver bullet for all trust issues? Blockchain-based trust patterns for collaborative business processes. In Int'l Conf. on Business Process Management (pp. 3–18). Springer, Cham.
54. Alshawi M, Salleh H (2013) IT/IS readiness maturity model. In Cases on Performance Measurement and Productivity Improvement: Technology Integration and Maturity (pp. 23–37). IGI Global.
55. Valdez-de-Leon O (2016) A digital maturity model for telecommunications service providers. Technology innovation management review, 6(8).
56. Kassab M, Destefanis G, DeFranco J, Pranav P (2021) Blockchain-Engineers Wanted: an Empirical Analysis on Required Skills, Education and Experience. In 4th Int'l Workshop on Emerging Trends in Software Engineering for Blockchain (pp. 49–55). IEEE.

57. Haque A, Alhashmi SM (2011) Parthiban R (2011) A survey of economic models in grid computing. Futur Gener Comput Syst 27(8):1056–1069
58. Chen Y (2020 Jun) Bellavitis C (2020) Blockchain disruption and decentralized finance: The rise of decentralized business models. J. of Business Venturing Insights. 1(13):e00151
59. Ante L (2020) Smart contracts on the blockchain–a bibliometric analysis and review. Telematics and Informatics:101519.
60. Bayhan S, Zubow A, Gawłowicz P, Wolisz A (2019) Smart contracts for spectrum sensing as a service. IEEE Trans. on Cognitive Communications and Networking, 5(3), 648–660.
61. Rodrigues B, Bocek T, Lareida A, Hausheer D, Rafati S, Stiller B (2017) A blockchain-based architecture for collaborative DDoS mitigation with smart contracts. In: IFIP Int'l Conf. on Autonomous Infrastructure, Management and Security (pp. 16–29). Springer, Cham
62. Xue J, Xu C, Zhang Y, Bai L (2018) DStore: a distributed cloud storage system based on smart contracts and blockchain. Int'l Conf. on Algorithms and Architectures for Parallel Processing (pp. 385–401). Springer, Cham
63. Javaid U, Aman MN, Sikdar B (2019) DrivMan: Driving trust management and data sharing in VANETS with blockchain and smart contracts. 89th Vehicular Technology Conf. (pp. 1–5). IEEE.
64. Almakhour M, Wehby A, Sliman L, Samhat AE, Mellouk A (2021). Smart Contract Based Solution for Secure Distributed SDN. 11th IFIP Int'l Conf. on New Technologies, Mobility and Security (NTMS) (pp. 1–6). IEEE.
65. Jain V, Kumar, B (2021) Combinatorial auction based multi-task resource allocation in fog environment using blockchain and smart contracts. Peer-to-Peer Networking and Applications, 1–19.
66. Gürer DW, Khan I, Ogier R, Keffer R (1996) An artificial intelligence approach to network fault management. Sri international. 86
67. Kumar GP, Venkataram P (1997) Artificial intelligence approaches to network management: recent advances and a survey. Comput Commun 20(15):1313–1322
68. Wang Y, Forbes R, Cavigioli C, Wang H, Gamelas A, Wade A, Strassner J, Cai S, Liu S (2018) Network management and orchestration using artificial intelligence: Overview of ETSI ENI. IEEE communications standards magazine. 2(4):58–65
69. Cayamcela ME, Lim W. (2018) Artificial intelligence in 5G technology: A survey. Int'l Conf. on Information and Communication Technology Convergence (pp. 860–865). IEEE
70. Wang CX, Di Renzo M, Stanczak S, Wang S, Larsson EG (2020) Artificial intelligence enabled wireless networking for 5G and beyond: Recent advances and future challenges. IEEE Wirel Commun 27(1):16–23
71. Gebremariam AA, Usman M, Qaraqe M (2019) Applications of artificial intelligence and machine learning in the area of SDN and NFV: A survey. 16th Int'l Multi-Conf. on Systems, Signals & Devices (pp. 545–549). IEEE
72. TMF, (2020) AI Management API Component User Suite, TMF 915
73. Singh S, Hosen AS, Yoon B (2020) Blockchain security attacks, challenges, and solutions for the future distributed IoT network. IEEE Access 9:13938–13959
74. Al-Breiki H, Rehman MH, Salah K, Svetinovic D (2020) Trustworthy blockchain oracles: Review, comparison, and open research challenges. IEEE Access 8:85675–85685

# AI-Enabled Intelligent Resource Management in 6G

Vijayakumar Ponnusamy and A. Vasuki

**Abstract** The next-generation 6G mobile networks are likely to be intelligent, highly dynamic, and extremely low latency to satisfy the needs of different diversified applications. With the increasing demand for wireless communications, resource management plays an essential role in providing higher data rates and extreme quality of service with the available resources. However, the complexity of allocating resources will become greater in the current ultra-dense heterogeneous infrastructures. With massive data and computing resources, the rapid progress of Artificial Intelligence (AI) eventually lightens the enormous capabilities needed for the future regularization of 6G and beyond. As a result, an AI-enabled network will be the most appropriate and suitable technique for intelligent resource management, automated network operations, and support in future complex 6G networks. This chapter will discuss the different machine learning techniques used for resource management in 6G networks, effective usage of available spectrum, prediction of the spectrum, and dynamic resource allocation.

**Keywords** Resource management · Massive connectivity · Network management

## 1 Essentials of AI in Resource Management

Development in wireless technologies still continues to satisfy demands for quick response time, higher bandwidth, and secure communication. All affordable physical layer technologies are utilized thoroughly after the development of post-3G systems in industries. This has led to the requirement of intelligent and optimized consumption

V. Ponnusamy (✉)
Department of ECE, SRM Institute of Science and Technology, SRM Nagar, Kattankulathur, Chengalpattu 603203, India
e-mail: vijayakp@srmist.edu.in

A. Vasuki
Department of ECE, SRM Institute of Science and Technology, Vadapalani campus, No.1. Jawaharlal Nehru Road, Vadapalani, Chennai 600026, India
e-mail: vasukia@srmist.edu.in

of the available resources. In this section, we discuss the importance of resource management and the role of AI in resource management toward achieving efficient next-generation 6G networks [1].

## 1.1 Challenges in Resource Management Toward Massive Connected Networks

Resource management aims to obtain proper employment of constrained physical sources to fulfill numerous traffic needs and enhance the machine's overall performance. The existing resource management techniques in academic usage are often developed for fixed networks. It depends upon the mathematical functions. In contrast, the conditions of the realistic wireless network are variable, which leads to regression of algorithms with high computational complexity. The application of assumed mathematical conditions of a static network in practical cases results in drastic performance loss. In addition to that, standard management schemes may be superior in extracting beneficial records related to users and networks. Various resource management techniques are desired for the massive number of nodes.

## 1.2 Resource Management in 5G

After the deployment of the 5G standards, academia and industries have started to focus on the next-generation wireless communication standard 6G. This technology achieves a high data rate of up to 1 Tb/s and a broadband frequency of 100 GHz to 3 THz [2]. Apart from important evaluation parameters of communication, Artificial Intelligence (AI) has been accepted as the essential feature of 6G by researchers in recent times. Wherein the applications of machine learning algorithms are believed to provide an appropriate solution for many complex scenarios. Network intelligence is expected to meet the challenges of heterogeneous networks. Though machine learning techniques have been employed in various applications, there is still scope for the realization of automated cellular communication systems. The existing problems in communication systems, architectures, and their performance should be focused on making use of 6G technologies.

## 1.3 Resource Management from 5G to 6G

The existence of AI can be seen in everyday scenarios. Nowadays, the amount of produced data by both machines and human is overwhelming, which exceeds the ability of humans to understand and digest that data and make decisions depending on that data. Thus, a hand of help from AI is needed to overcome such challenges. The 5G LTE communication system is a promising solution to provide a high user experience in terms of the provided speed, amount of data, and cost. However, due to its complexity, the technology of LTE needs some improvement in terms of resource management and optimization. With the aid of AI, these two challenges can be overcome. The AI represented by the improved Q-learning algorithm with the Self-Organizing Network (SON) concept in LTE is used to manage and optimize Handover (HO) parameters and processes in the system [3].

AI has become necessary in day-to-day activities. The data produced by humans and machines is huge. There is a need for humans to understand the generated data and analyze it. This helps in making appropriate decisions. This can be easily solved by employing AI. The 5G LTE in communication has been considered as an auspicious solution in terms of speed, data rate, and cost. However, there is a need for some enhancements in resource management and optimization techniques which could be overcome by using AI techniques. AI-aided Q-learning algorithm along with Self-Organizing Network (SON) in LTE manages and optimizes the handover parameters and different stages in the system [4]. The challenges in AI-based 6G networks are depicted in Fig. 1. The role of AI-based algorithms is illustrated in Fig. 2.

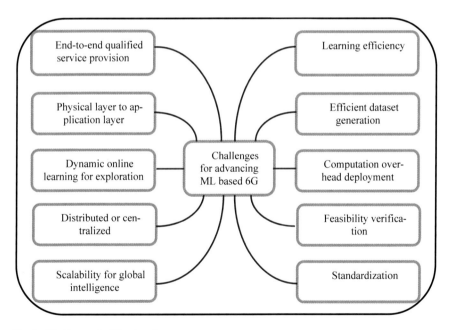

**Fig. 1** Challenges in 6G networks

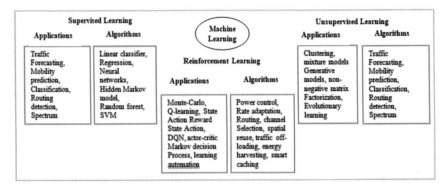

**Fig. 2** Role of AI algorithms in wireless networks

## 2 Application of Machine Learning Techniques in Network Management

To fulfill the enormous requesting management, the growth of 6G has been considered as progress that overcomes the limits of improved broadband, unlimited access, and ultra-reliable latent management in 5G remote networks. As of late, the construction of 6G networks has been amazingly heterogeneous, thickly conveyed, and dynamic. In order to reach the best tight Quality of Service (QoS), this complex structure will lead to a defect in network activity routines. Therefore, machine learning is emerging as a crucial answer to acknowledge completely wise organization coordination and the executives. By gaining from unsure and dynamic conditions, machine learning-based medium assessment and bandwidth management will allow the executives to open up promising circumstances for bringing the amazing presentation of ultra-broadband strategies, like terahertz interchanges. Furthermore, difficulties carried by ultra-massive networks concerning power and end-to-end secure communication can be moderated by employing appropriate machine learning-based techniques. Also, shrewd versatility among the executives and asset assignment will ensure the ultra-unwavering quality and low inactivity of administrations. Concerning these issues, this chapter presents and studies some cutting-edge methods dependent on machine learning and their applications in 6G to help with ultra-broadband, ultra-massive access, and low inertness management [5].

### 2.1 ML-Enabled Broadband Transmission in 6G Networks

The available spectrum is scarcely enough to satisfy the expanding needs. For example, some arising applications, like holography, may require a data rate of up Tb/sec, which is roughly three times of magnitude faster compared to average 5G correspondences. Accordingly, THz correspondences, using groups within the scope of 0.1–10 THz such as 140, 220, and 340 GHz frequencies, are required to help an information pace of up to terabits each second. To accomplish such a limit moving toward execution, exact data of time-shifting channels is particularly imperative to advance the terahertz transfer speed portion and further develop range productivity. In this segment, we present some cutting-edge AI/ML applications in terahertz channel assessment and range the executives [6].

At the terahertz recurrence groups, the channels experience the ill effects of high climatic retention coming about because of the water fume noticeable all around, which impacts misfortunes fundamentally. Moreover, the free-space path loss is likewise unavoidable actually as far as climatic constriction is concerned. Moreover, terahertz channels are seen as non-fixed, particularly for dynamic situations where the two clients and items may be moving. Accordingly, customary channel models dependent on suppositions of being fixed or semi-fixed can't currently have any significant bearing on the ultra-wideband spectrum. Machine learning calculations are equipped to examine the correspondence information and foresee likely signs of misfortune in a guaranteed or obscure climate. Subsequently, a wide range of Artificial Intelligence calculations could be employed to the physical layer of 6G organizations that manage the hardships portrayed for ultra-wideband spectrum assessment.

In many applications, in order to further develop assessment precision in powerful situations, the Bayesian filter based on reinforcement learning has been acquainted with the direction of arrival assessment in THz directs in present examinations. In particular, the Bayesian channel executes the assessment of the current direction of arrival from both current estimation and past gauges. In this methodology, the earlier change probabilities between framework states are essential to the assessment execution of the Bayesian filter. A reinforcement learning algorithm could be employed to enhance the probabilities of state changes from the results of previous measures and henceforth work on the exhibition of the Bayesian filter. The major classifications and applications of machine learning algorithms are described below.

- The learning algorithm which is based on enough and efficient training with labeled data is known as *supervised learning*. These algorithms can be acquainted with transmission path loss and shadowing forecast, obstruction board, channel assessment, etc. Some of the architectures, such as Deep Neural Networks (DNN), K-Nearest Neighbor (KNN), and Support Vector Machines (SVM), are based on supervised learning algorithms.

  KNN algorithm is used for identifying and allocating the resources (spectrum/power/positioning) requirements. For example, as shown in Fig. 3, the new requirement can be categorized using KNN algorithm, where $K = 3$ represents the distance from the new sample (data) is calculated with three nearest data points. KNN uses Euclidean distance (d) for classification which is expressed in (1)

$$d = \sqrt{(x_2 - x_1)^2 + (y_2 - y_1)^2} \tag{1}$$

**Fig. 3** An example of KNN

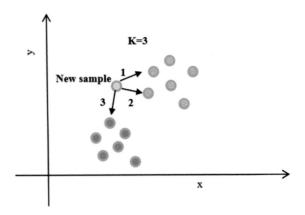

Similarly, SVM also is used to solve handover problems in heterogeneous networks. Antenna selection in massive MIMO is achieved with the aid of an SVM classifier among hundreds of antennas. Estimating channel noise for effective network management.

- In any wireless communication network, channel estimation and modeling play an essential role in determining the end-to-end performance of the communication system. These problems could be easily solved by yet another category of machine learning called unsupervised learning mechanisms. *Unsupervised learning* does not require training with labeled data. The major applications are suppressing interference, user clustering, and overcoming challenges in duplex mode. K-means clustering and fuzzy-C-means algorithms are some of the most popular unsupervised learning mechanisms in recent communication technologies.

The Multipath Component (MPC) analysis is an essential task in gathering information on the wireless channel. K-means clustering algorithm forms the grouping in the MPC to reduce the Euclidean distance between the data until it becomes converging. It groups the parameters such as the delay (t), azimuth angle of arrival (AoA), azimuth angle of departure (AoD), the elevation angle of arrival (EoA), and the elevation angle of departure (EoD) which have similar behavior. The following example illustrates the step by execution of K-means clustering:

Step1: It starts initial clusters randomly with centers ($k_1$, $k_2$, $k_3$), as shown in Fig. 4a.

Step 2: It assigns each data point closest cluster center, as shown in Fig. 4b.

Step 3: Move each cluster center to the mean of each cluster as shown in Fig. 4c.

Step 4: It calculates the clusters related to the new data points as shown in Fig. 4d.

Step 5: Recomputes the cluster centers as shown in Fig. 4e.

Step 6: Move the cluster centers to cluster means as shown in Fig. 4f.

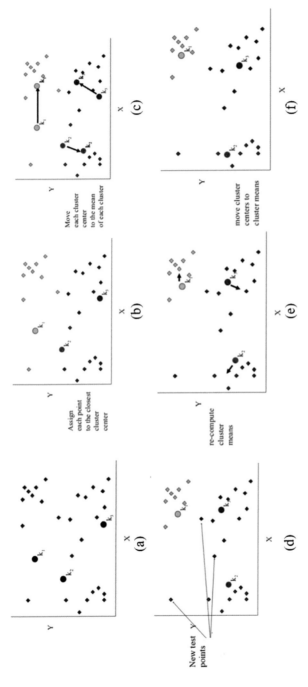

**Fig. 4** An illustration of K-means clustering algorithm in unknown data points

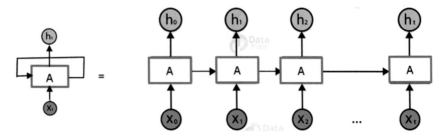

**Fig. 5** An LSTM architecture

- Another important category is *Deep learning.* Deep learning algorithms are employed for extracting channel characteristics, finding dynamic channel information, detecting different modulated signals/symbols, and recovering the original raw input from encoded data. The major deep learning structures such as Convolutional Neural Network (CNN), Recurrent Neural Network, Deep feed Forward Neural Network, and Deep Belief Networks (DBN) could be employed depending upon the applications.

CNN could be applied for dynamic power control for improving Non-Line of Sight (NLOS) transmission, automatic modulation classification, and channel estimation in a next-generation wireless communication system. Localization in wireless networks is a challenging task because of shadowing and multi-path fading in indoor environments. In order to fulfill the requirements of 6G wireless communication, the THz spectrum provides ultra-wideband for application. An LSTM architecture (as illustrated in Fig. 5) could explore the channel state information of THz wireless signals, AoA, received power, and delay which are useful parameters of the indoor environment. A series of temporal data $(X_o, X_1, \ldots X_t)$ is given input to the LSTM.

- *Reinforcement learning* is a combination of supervised and unsupervised learning algorithms. These algorithms are based on action rewards. The agent in the application continuously gets the information from the environment to take appropriate action. If the action is successful, the algorithm gets positive feedback to further explore the new unknown state. Reinforcement learning-based algorithms are mainly used in tracking the unknown wireless medium and choosing feasible modulation modes. Markov decision process, Q-learning, and fuzzy-reinforcement learning are algorithms used often.

Markov decision process applies probability to predict the future requirements based on the present observed state in the channel estimation process which is illustrated in Fig. 6.

The application of the deep learning/machine learning algorithm in various levels of the 6G network is illustrated in Fig. 7.

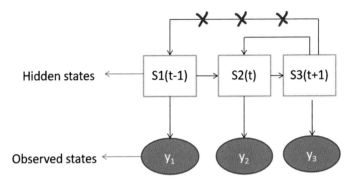

**Fig. 6** An example of the Markov process

## 3 Application of Machine Learning Techniques in Network Management

In this section, the necessity of machine learning in a massive connected network, spectrum prediction based on ML will be discussed [7].

### 3.1 Need for Massive Connectivity Management in 6G Network

In high-dimensional networks, connectivity among different devices can allow the terminals to create diversified integration with various levels of network and make the network to reach uplift in coverage. Because of the dynamic nature of the connectivity, there is a chance for inserting and removing devices to cope with massive connectivity effectively. The connectivity between network and physical layer bears with high delay in higher order networks. This delay results in data transfer, which could be optimized in their end-to-end connectivity. Managing connectivity among numerous devices depends on present wireless channel conditions, wherein the portability of Base Stations (BSs) makes getting channel state information more tedious.

### 3.2 Radio Resource Management in High-Dimensional Networks

In massively connected networks, coverage of indistinguishable areas is done by allowing many devices to the network. This results in severe interference in the network, which could be suppressed by using a scheduling mechanism. This technique provides information about the dynamic channel to improve scheduling among

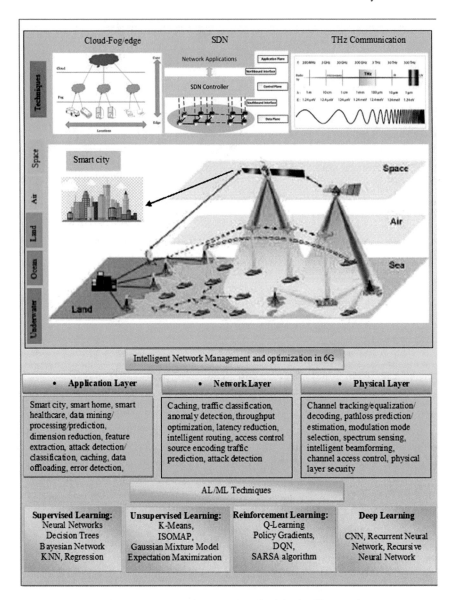

**Fig. 7** Application of deep learning/machine learning algorithm in 6G network

users. In contrast to hyperlinks between terrestrial BSs, the inter-satellite, inter-aerial, inter-satellite–aerial, and inter-satellite–terrestrial are non-ideal, and hence has a poor delay. So, there is a necessity for collaborative scheduling in massively connected networks. Moreover, long-distance and high-mobility shifting BSs lead to the severity in channel approximation, which makes designing the feasible algorithm as a complex one.

## 3.3 AI-Enabled Automated 6G Networks

The key enables methods used in 5G networks such as software-defined networks; network function virtualization is considered in 6G networks also. These techniques have the capability of making self-defined networks that are suitable for 6G networks. Anyhow, these networks should be introduced with intelligence in the case of 6G networks. With the integration of artificial intelligence in wireless networks, intelligence could be achieved with sufficient training. So, machine learning/deep learning/Artificial Intelligence has been considered as the most suitable technique for developing 6G networks as self-defined networks.

The integration of Artificial Intelligence in software-defined networks can obtain effective network architecture. Good optimization applied in 6G makes it autonomous networks compared to 5G networks.

Machine learning-based network management has the ability to organize the network architecture, slice and collect different channel access technologies to get smooth networks, and satisfy the requirements of dynamic services and applications. The network performance metrics are continuously observed with the aid of AI-based optimization techniques. Based on the observations, network parameters could be updated to maintain the best services. These AI-based techniques can get practical network performance, which is considered as historical data. Network intelligence can also be obtained by deploying multilevel AI. In huge connected networks, AI could be integrated at the routers to forward the data. AI-enabled techniques were introduced in Radio Access Networks to manage multiple BS-related processes like portability and interference suppression. AI will enable the large IoT network to shift from the facts center to the edge of the community.

## 4 Application of Machine Learning Techniques in Network Management

In this section, we will discuss how machine learning/deep learning techniques are used for allocating the available resources in terms of multi-access edge computing, mobility, handover management, and spectrum management [8].

## 4.1 Deep Learning-Based Multi-Access Edge Computing

Multi-access edge computing has become inevitable in the 6G networks. MEC furnishes mathematical and analytical computations and brings radio access networks to mere closeness with other devices. Because of their multidimensional, randomly uncertain, and dynamic properties, MEC's firm-level maximization, whereabouts, and sample training are important aspects. Hence, a conventional Lagrangian duality algorithm may face threads under complex networks. AI-enabled strategies can retrieve past records to know and assist for optimization, forecasting, and choice in MEC. Figure 4 llustrates the architecture of AI-enabled cell edge computing, cloud and edge computing. As proven because of the confined capability, lightweight AI algorithms may be applied to offer clever programs for other situations like transportation and agriculture. For instance, reinforcement learning-based edge computing is helpful in resource management and is considered a model-free scheme since it is not essential to have any prerequisite knowledge about any area. This can be implemented in analyzing the changes in the environment and choosing the appropriate method in practice.

- Extensive deep learning algorithms are expected to furnish functional training on the central cloud server. The AI-enabled classification algorithm is utilized to optimize traffic flow decisions for different service applications that are dynamic and different. An AI-based cluster is used to reduce complexity in MEC servers instead of individual decisions.
- The data received from the edge computing servers can be well trained to extract inherent features automatically. In such cases, the deep learning algorithm could be suited to train the system model to obtain the type of service, traffic, and security for which it is designed.
- In addition to that, Deep Reinforcement Learning (DRL) is made active in finding adequate resource management policy in extremely complex and time-varying MEC networks. To map the decisions and the environment, optimal resource management policies are used. Edge devices are supported with high-quality services by adopting DRL to use historical knowledge by improving efficiency and accuracy.

## 4.2  AI-Based Handover Management

Being 6G networks are highly discrete, multi-layer and large-dimensional, mobility, and handover management are highly challenging issues of 6G networks. Mobility prediction and handover are achieved by applying suitable AI techniques to achieve uninterrupted services.

*Role of AI in UAV networks:* In the case of integrating UAV communication with 6G networks, this results in frequent handovers. Suppose, if UAV communications are combined with 6G networks, and the fast dynamic mobility of UAVs results in established handovers. Furthermore, the subject is extended in processing efficient handover due to the numerous provider needs such as high statistics, high trustability, and low latency. The immoderate mobility of devices and UAVs results in unpredictability about their locations. This problem could be resolved by one of the AI techniques, reinforcement learning. It learns to optimize the handover techniques in the practical scenario to explore the mobility behaviors of devices/UAVs in online mode. This approach reduces the transmission delay and ensures reliable Wi-Fi connectivity.

Figure 4 shows the factors of shrewd mobility and handover management based on DRL in UAV-enabled networks. In these networks, every UAV can act as an agent to interact with the environment, thereby enhancing the network coverage. Each agent observes the nearby states and finds the most appropriate action to achieve the positive reward. The reward can be manipulated with the help of connectivity, delay, and feedback from the environment. DRL-based UAV network can have the ability to process handover mechanically and reduce the delay, handover failure probability. At the same time, it offers the best grade of service.

*Role of AI in Vehicular networks:* Large-scale vehicular networks should have ultra-high speed and low delay insensitive with the evolution of 6G networks. Therefore, efficient mobility management is an essential parameter to achieve end-to-end reliable and low latent communication. The mobility patterns of high-speed vehicular users are continuously learned by the popular deep learning structures like RNN and ANN. These AI-enabled techniques successfully mitigate conventional handovers, handover failures. Similarly, another deep learning model called Long Short-Term-Memory (LSTM) is utilized for solving handover problems, where it exploits the history of past and future mobility and provides the prediction of sequence, trajectories of vehicles to optimize the handover parameters.

A multilayered LSTM integrated with autoencoder is used to resolve the frequent handover problems in wireless communication.

## 4.3  AI-Based Spectrum Management

The evolution of 6G networks requires exceptional frequency spectrum such as low RF, millimeter wave, THz, and visible band to provide excessive information data rates, as illustrated in Fig. 4 [9, 10]. When a broad range of gadgets are used in 6G networks, it leads to spectrum management. AI enabled is considered a successful technique for making massive connectivities among devices, as shown in Fig. 4.

- In general, AI-based structures have three layers, namely, input layer, hidden layer, and output layer. A wide range of spectrum datasets is given as input. Then it trains the hidden layers to find out considerable characteristics of spectrum utilization. At last, the effective spectrum management methods are derived in the output layer in the practical scenario to support numerous connectivity among devices.
- The proper training of the AI framework forces an offline training model to easily recognize online spectrum management solutions. Various spectrum bands could be better used for making remarkable transmissions with massive bandwidth, wherein low-frequency bands could be allocated for broadcasting information for satellite-ground transmissions.

# 5   AI-Enabled Dynamic Resource Allocation

In any wireless network, spectrum utilization, process computation, and architectures are crucial resources beyond 6G. Therefore, the enhancement of useful resource allocation is achieved by introducing dynamic resource allocation methods. Dynamic resource allocation is implemented with the aid of AI and blockchain technologies. AI-based dynamic resource utilization is discussed in this section [11–16].

## 5.1   *Efficient Sharing of Radio Sources*

In wireless communication systems, effective usage of available radio resources is an essential process to improve uninterrupted service providing for users. Radio resource allocation methods are restricted based on various network parameters. For example, the uncertainty caused in information transmission at Wi-Fi networks subsequently affects the data rate and processing time. Dynamic, varying channel parameter creates interference, which is because of high traffic and mobility in the environment. Radio resource allocation has become a highly challenging task when forecasting the services for different kinds of users with a limited spectrum.

*Categories of resource allocation:* Effective allocation of radio resources is classified as centralized or decentralized.

*Centralized method:* These methods are processed mainly based on a single core entity that gathers information from the users of the wireless networks. After that, resources are allocated based on the capabilities of the network. These methods provide outstanding responses with compromised data transfer.

*Decentralized method:* In the case of decentralized techniques, users are permitted to make selections on their own. These techniques are more flexible than centralized with the aid of sub-optimal results.

The above-discussed resource allocation schemes change based on the optimization techniques used for achieving throughput enhancement, processing delay, user fairness, energy, and spectral efficiency. So, the effective resource allocations are categorized based on the type of networks, optimization technique, and proposed machine learning algorithms.

Achieving optimal results in radio resource allocation is a tedious objective due to the requirement of various parameters. This could be solved by the heuristic approach, which loosens up the network expectations and finds a sensible alternate solution. But these methods are not promised to obtain excellent results. The other approach is theory-based game techniques; the network nodes are considered players are interrelating and influencing the other's options. Each player has a couple of options to maximize utility. The main advantage of game theory approaches is the flexibility to alter based on network dynamics.

## 5.2  AI-Based Dynamic Spectrum Allocation

Dynamic spectrum allocation could be solved by multi-agent deep reinforcement learning methods, wherein each user occupancy in the spectrum is referred to as the agent. In this approach, a multi-agent environment is considered a Markov game model. A Neighbor-Agent-Actor-Critic (NAAC) model is proposed to tackle the uncertainties in the environment. This model trains the neighbor nodes from the statistics in a centralized manner. The relationship among devices that share the spectrum at the same instance improves device performance like achievable data rate and spectrum efficiency. Reinforcement learning is used in such a network for training the nodes based on past records.

## 5.3  AI-Based Distributed Spectrum Access

Distributed dynamic resource access could be developed for common real-time networks effectively. At the same time, the computational consumption of huge networks and incomplete observations in the network should be eliminated. LSTM approach is employed to achieve the mentioned objective, which has an internal layer with combinations of previously measured values. A dual Deep-Q-Network (DQN) is also employed to reach expected rewards from unknown states. Users need to use their trained DQN weight values by communicating with the central unit and then map its local commentary to spectrum get entry to move based on the learned DQN.

## 6  AI-Enabled Dynamic Resource Allocation

As of now, there is no limitation on terahertz frequency band utilization. The spectra play an important role in some different applications, like satellite services, spectroscopy, and meteorology. As of late, the Federal Communications Commission has been putting resources into using terahertz ranges for portable administrations and applications. Accordingly, range sharing strategies are important in conjunction with future terahertz interchanges and the other applications recorded beforehand. What's more, as examined in the past area, 6G networks will, in general, be multidimensional, super thick, and heterogeneous. Consequently, taking into account that the engendering medium and divert trademark in coordinated 6G networks are altogether unmistakable contrasted and earthbound organizations in 5G, it requires more exertion to upgrade the range of the executives of terahertz interchanges in 6G [17–20].

Reinforcement learning can possibly acknowledge the keen or astute range of the board to manage these issues, particularly when a lot of information can be utilized to prepare and anticipate. These preparation and forecast results can be exploited to settle on choices concerning whether the bandwidth is involved and to make a move, for example, getting to or delivering the range band. What's more, through the collaboration among clients and the remote climate, clients can enhance their methodologies iteratively to amplify the worth of remuneration capacities, which can be set up thinking about range productivity, network limit, devoured energy, obstruction, etc. Notwithstanding, reinforcement learning isn't skilled for learning a viable activity esteem strategy when there exist arbitrary commotion or estimation blunders involving the state perceptions, implying that the quantity of states within sight of irregular clamor is endless by and by. Resolving this issue of arbitrary state estimations, deep reinforcement learning is considered as an appropriate method to solve the streamline options in 6G networks, which incorporate flexible spectrum access, transmit power optimization, and radio spectrum allocation.

## 6.1  Reinforcement Learning-Enabled 6G Network

Recently, intensive analysis on developing Beyond 5G networks has been developed, moreover delivered up into 6G wireless networks which have geared toward transportation, ultra-reliable, low-latency conversation services. Reinforcement learning-based algorithms are adapted to obtain the possible resources from the environment to optimize the capacity in heterogeneous networks. For instance, if deep learning algorithms are used to improve performance, it requires huge memory for storing the data and high computational tasks. In the case of the reinforcement learning mechanism, the agent or device understands the actions to be taken in order to maximize the reward for the related actions. Therefore, a reinforcement learning-based algorithm learns the possible protocols and actions to match the recent dynamic unknown states. These algorithms create a new environment with states, actions, rewards, and state-action transition probabilities. It has been proven that reinforcement learning algorithms are more suitable for future generation networks.

*Basics RL-based approach:* The conventional interaction between agent and environment is illustrated in Fig. 6. The standard reinforcement learning algorithm has three terms such as (i) Policy, (ii) Reward, and (iii) States. *Policy:* Policy is an essential part of reinforcement learning algorithm; it defines the method to describe about how an agent could interact with the environment. *Reward:* For every action, the agent gains reward/feedback from the system. Based on the value obtained as a reward, decide the possible state-action policy for that particular application/environment. *Q-function:* It describes how long the algorithm earns a reward for the action taken. The reward for the accurate action must be small, but it would be considered as a worthy Q-value for long-haul operation (Figs. 8 and 9).

Q-value for the action taken is given

$$Q(s, a) = r(s, a) + \gamma \max Q(s', a) \tag{2}$$

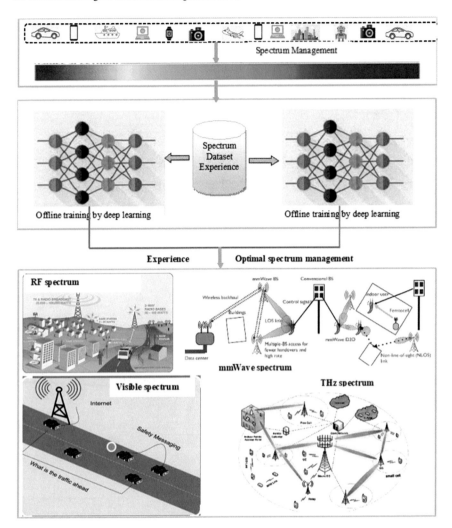

**Fig. 8** Deep learning-based spectrum management

The expression (2) states that Q-value achieved from being at state s and taking an action a, with the reward of r(s,a) plus the possible Q-value from the next state s'. The typical Q-learning and deep Q-network is illustrated as shown in Fig. 10.

Suppose the high-value action is taken very often with the help of the agent results in exploitation in unknown surroundings. Q-learning must be a part of all model-free reinforcement learning algorithms to solve the channel behavioral problems. It sets use of learning rate to change the ability of learning, bargain problem to carry higher/lower well worth to the long-run reward, immediate reward, and update in

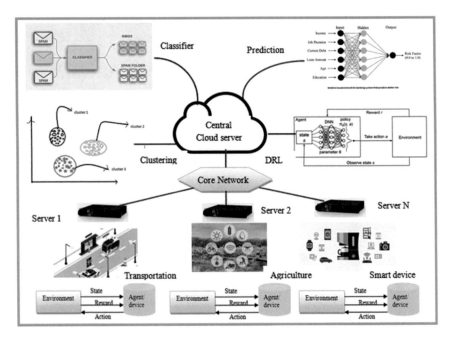

**Fig. 9** Deep learning-based edge computing architecture

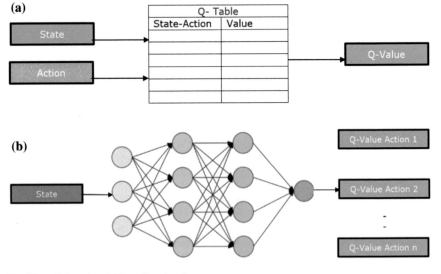

**Fig. 10** **a** Q-learning. **b** Deep-Q-network

Q-value function to replace present Q-value perform. Most appropriate action decisions maximize the Q-value in the proposed algorithm. Deep-Q-learning methods are mostly used in cognitive radios and channel access in Wi-Fi networks.

## 6.2 Realization of RL-Based Framework for 6G Networks

In this section, the approximation of reinforcement learning algorithms to realize the channel state estimation for WLAN networks is discussed. A hybrid RL aware framework is proposed, wherein the main module executes the action plans (policy). *Learning phase:* The algorithm/model learns the channel behaviors from the stored channel state (data). *Exploitation:* The model optimizes the useful resources based on the training. Figure 11 illustrates the important layers of reinforcement learning-based architecture for WLANs in mIoT (mobile Internet of Things) environment. The model/system is trained at the AP to collect the channel behavior statistics from many state actions. The model placement is also performed to get quick responses to circumstance actions. The model could be re-trained again and again based on the updated measured records.

*Training phase:*

In the reinforcement learning framework, the model in a Wi-Fi environment learns the channel for state transition, which avoids possible collisions. Then, the BS collects the information of various agents/models during the uplink transmission. The probability of collision could be used for learning and algorithm to help the Medium Access Control (MAC)-Resource Allocation layer during frequency band selection. Then the gathered record is pre-processed with the reinforcement learning technique to learn the medium accurately. For example, the application of Q-learning transforms the gathered information as a reward value. Based on the application, the reward can be scaled between 0 and 1. Even as developing the reinforcement learning-based model, protocols have to be considered. For instance, based on available resources, an AP may require the widest variety of related STAs. The policies are strongly related to the competencies of the Wi-Fi gadgets. Once the reinforcement learning algorithm on the BS produces the output, it's far disbursed all over the network environments to the STAs, which can be then arranged/remodeled to provide fast ideal spectrum aid allocation to new requests.

*Placement phase:*

In this phase, BS can allocate a new spectrum for the requests or hand over the spectrum based on the stored behavior from state actions. BS processes the collected records as a part of the learning phase. The well-known Q-learning algorithm is applied on the AP to provide a reward-based response for future requests. The channel access or allocation status is communicated to the corresponding state actions.

## 6.3 Reinforcement Learning-Based Spectrum Management

There are no restrictions on using the THz spectrum. Satellite-TV link, spectroscopy, and meteorology occupied the THz spectrum already. Nowadays, cellular services and applications are also allowed to use the THz band. So, spectrum allocation and sharing methods proposed for sub-GHz [21–23] are essential in THz communications. Based on the discussion about 6G networks in the previous sections, future generation networks are high-dimensional, ultra-high density, and heterogeneous networks. Therefore, 6G networks should have better propagation medium and channel state estimation compared with conventional 5G networks. That is, more efforts should be involved in achieving spectrum management in THz communications.

- Reinforcement learning algorithms have greater capabilities to solve the spectrum management problems in THz communication, though it involves a large amount of the previous history of records for learning. These learning and prediction mechanism results in high advantage while taking decisions for action. The actions find whether the spectrum is already occupied or free to further access/allocate the spectrum.
- The continuous interaction between users and the wireless environment makes to optimize the proposed technique for the achievement of maximum reward values. The reward values are referred to spectrum efficiency, networkability, and interference mitigation and received signal strength depending upon the applications. Reinforcement learning also has a significant state-action-reward policy, which could process precisely though there is a constant presence of randomness in channel noise. In order to address these problems, deep reinforcement learning is utilized as an appropriate method for spectrum management of massively connected networks (Figs. 11 and 12).

## 7  Conclusions

This chapter discusses the importance of resource management in Ultra-Reliable Low-Latent Communication (URLLC). The challenges and problems in the high-dimensional networks can be resolved by utilizing suitable machine learning/deep learning algorithms. We discussed the opportunities in terms of essentials of AI in resource management, applications of machine learning techniques in network management, machine learning-based spectrum prediction, efficient utilization of resources, dynamic resource allocation, and the role of reinforcement learning in solving problems.

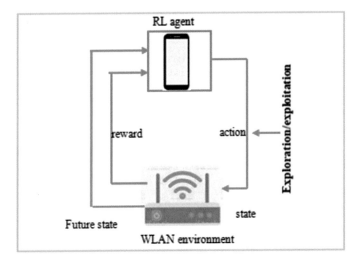

**Fig. 11** An example of deep reinforcement learning

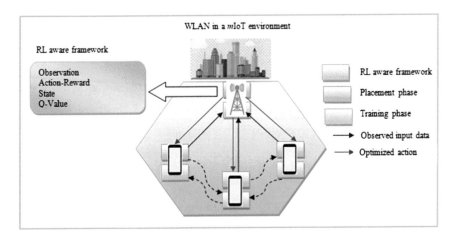

**Fig. 12** Key stages of RL-based WLAN in *m*IoT environment

## References

1. Shen Y, Shi Y, Zhang J, Letaief KB (2020) LORM: learning to optimize for resource management in wireless networks with few training samples. IEEE Trans Wireless Commun 19(1):665–679
2. Mardan HA, Ahmed SK (2020) Using AI in wireless communication system for resource management and optimization. Period Eng Nat Sci
3. Sun Y, Peng M, Zhou Y, Huang Y, Mao S (2019) Application of machine learning in wireless networks: key techniques and open issues. IEEE Commun Surveys Tutorials, pp 3072–3108
4. Kato N, Mao B, Tang F, Kawamoto Y, Liu J (2020) Ten challenges in advancing machine learning technologies toward 6G. IEEE Wireless Commun pp 96–103.

5. Ali R, Ashraf I, Bashir AK, Zikria YB (2021) Reinforcement-Learning-Enabled Massive Internet of Things for 6G Wireless Communications. In IEEE Communications Standards Magazine, pp 126–131
6. Du J, Jiang C, Zhang H, Wang X, Ren Y, Debbah M (2020) Machine learning for 6G wireless networks: carrying forward enhanced bandwidth. Massive Access, and Ultrareliable/Low-Latency Service, IEEE Vehicular Technology Magazine, pp 122–134
7. Zhang Z et al (2019) 6G Wireless Networks: Vision, Requirements, Architecture, and Key Technologies. IEEE Vehicular Technology Magazine, pp. 28–41
8. Yang H, Alphones A, Xiong Z, Niyato D, Zhao J, Wu K (2020) Artificial-intelligence-enabled intelligent 6G networks. IEEE Netw: 272–280
9. Lin K, Li Y, Zhang Q, Fortino G (2021) AI-Driven collaborative resource allocation for task execution in 6G-enabled massive IoT. IEEE Internet of Things J 5264–5273
10. Chen Y, Liu W, Niu Z, Feng Z, Hu Q, Jiang T (2020) Pervasive intelligent endogenous 6G wireless systems: prospects, theories and key technologies. Digital Commun Netw
11. Ahokangas P, Matinmikko-Blue M, Yrjölä S, Hämmäinen H (2020) Spectrum management in the 6G Era: the role of regulation and spectrum sharing.In: 2020 2nd 6G wireless summit (6G SUMMIT), pp 1–5
12. Nawaz SJ, Sharma SK, Wyne S, Patwary MN, Asaduzzaman M (2019) Quantum machine learning for 6G communication networks: state-of-the-art and vision for the future. IEEE Access 46317–46350
13. Gür G (2020) Expansive networks: Exploiting spectrum sharing for capacity boost and 6G vision. J Commun Netw pp. 444–454.
14. Ali et al (2020) 6G White Paper on Machine Learning in Wireless Communication Networks
15. Hu S, Liang YC, Xiong Z, Niyato N (2021) Blockchain and artificial intelligence for dynamic resource sharing in 6G and beyond. IEEE Wireless Commun
16. Guan W, Zhang H, Leung VC (2020) Customized slicing for 6G: enforcing artificial intelligence on resource management. IEEE Netw
17. Sodhro AH et al.(2020) Towards ML-based energy-efficient mechanism for 6G enabled industrial network in box systems. IEEE Trans Indust Informat
18. Naderializadeh N, Sydir JJ, Simsek M, Nikopour H (2021) Resource management in wireless networks via multi-agent deep reinforcement learning. IEEE Trans Wireless Commun: 3507–3523
19. Peng H, Shen X (2021) Multi-agent reinforcement learning based resource management in MEC- and UAV-assisted vehicular networks. IEEE J Selected Areas Commun: 131–141
20. Feriani A, Hossain E (2021) Single and multi-agent deep reinforcement learning for ai-enabled wireless networks: a tutorial. IEEE Commun Surveys Tutorials, 1226–1252
21. Vijayakumar P, Malarvihi S (2017) Green spectrum sharing: genetic algorithm based SDR implementation. Wireless Pers Commun 94(4):2303–2324
22. Vijayakumar P, Malarvizhi S (2016) Reconfigurable filter bank multicarrier modulation for cognitive radio spectrum sharing-a SDR implementation. Indian J Sci Technol 9(6):1–6
23. Ponnusamy V, Malarvihi S (2017) Hardware impairment detection and prewhitening on MIMO precoder for spectrum sharing. Wireless Pers Commun 96(1):1557–1576

# Role of Blockchain and AI in Security and Privacy of 6G

Hany F. Atlam, Muhammad Ajmal Azad, Manar Altamimi, and Nawfal Fadhel

**Abstract** In the coming era, 6G is expected to bring a new reality that contains billions of things, humans, connected cars, robots and drones that will produce Zettabytes of digital data. 6G is mainly used to design an inclusive digital and physical environment, which is able to sense it, understand it and programme it. Several countries around the world are competing to own 6G infrastructures and solutions since this new technology provides huge capabilities that will reshape how enterprises operate. Although 6G provides several advantages over existing technologies, security and privacy issues still need to be addressed. This is because 6G provides automatization of most critical processes, which produce a more wide and complex attack surface. Also, with 6G, the network becomes more vulnerable not only to direct security attacks but also to misbehaviour of automated processes that require to be recognized, and their effect should be minimized. This chapter provides a discussion of security and privacy issues in 6G and how the integration of blockchain and Artificial Intelligence (AI) with 6G can provide possible solutions to overcome these issues. The chapter starts by providing an overview of wireless communications technologies from 0 to 6G. This is followed by discussing the main security and privacy issues in 6G networks. Then, the integration of blockchain with 6G will be discussed by highlighting possible solutions to overcome security and privacy issues associated with 6G. The integration of 6G with AI will be also discussed by highlighting

H. F. Atlam (✉) · M. A. Azad
Department of Engineering and Technology, University of Derby, Derby D22 1GB, UK
e-mail: h.atlam@derby.ac.uk

M. A. Azad
e-mail: m.azadh.atlam@derby.ac.uk

H. F. Atlam
Computer Science and Engineering Dept, Faculty of Electronic Engineering, Menoufia University, Menoufia, Egypt

M. Altamimi · N. Fadhel
School of Electronics and Computer Science, University of Southampton, Southampton, UK
e-mail: m.m.m.altamimi@soton.ac.uk

N. Fadhel
e-mail: Nawfal@soton.ac.uk

the importance of AI in 6G and how AI with 6G can provide better and effective security and privacy solutions. In the end, healthcare with 6G is presented as a use case by highlighting security issues and discussing the role of AI and blockchain in providing effective security solutions in the healthcare sector.

**Keywords** 6G · Security · Privacy · Security and privacy · Blockchain · AI · Blockchain with 6G · AI with 6G

## 1  Introduction

Although the fifth generation or 5G of mobile communication network is not fully implemented, several studies started to talk about 6G (Sixth Generation) and its potentials. 6G is anticipated to enable unprecedented Internet of Everything (IoE) applications with enormously varied and challenging demands. 6G is intended as a space-aerial-terrestrial-ocean interconnected three-dimension network with multiple sorts of parts enabled by new technologies and standards to make the system more intelligent and flexible to meet varied requirements efficiently [1]. Some people argue that 6G networks will be just a faster version of 5G, but the reality is 6G, is a new improved version of 5G in almost all aspects. For instance, unlike the 5G network, coverage will not be confined to the ground level. Instead, it should cover the entire undersea surface area. Artificial Intelligence (AI) capabilities also will be substantially higher on the 6G network [2].

6G provides new attractive characteristics that will integrate capabilities of previous mobile communication technologies such as high reliability, massive connectivity, high throughput and network densification. 6G can be able to provide several benefits to various applications such as autonomous vehicles, sensing, implants, computing reality devices, smart wearables, and 3D mapping [3, 4]. 6G will also improve performance and maximize user Quality of Service (QoS). 6G is predicted to offer 1000 times faster wireless connectivity than 5G. Additionally, 6G is projected to enable ultra-long-range communication with a latency of less than 1 ms [5].

Although 6G provides unlimited advantages for various applications due to the huge capabilities it provides, security and privacy are the major issues that 6G need to address. The integration of AI in 6G can be used to develop safer and secure systems, but it can also be used to create more dangerous attacks. Physical layer security measures can also be used as the first line of defence for safeguarding network portions that haven't been thoroughly explored. Therefore, there is a need for effective solutions to security and privacy challenges in 6G networks.

Adopting one of the Distributed Ledger Technologies (DLTs) such as blockchain is one of the solutions for overcoming security and privacy challenges in 6G. Blockchain is distributed, decentralized, and trustless network by eliminating the centralized authority. This improves transparency by sharing the transaction details between the participants in the network, provides rigorous security to prevent cyber-attacks such as Denial of Service (DoS) and privacy to protect sensitive information and prevent data manipulation [6]. Blockchain is also capable of storing data across the network immutably and securely. This eliminates the single point of failure and data manipulation [7].

Another technology that can provide an effective solution for security and privacy issues in 6G is AI. The integration of AI with 6G will bring several benefits for overcoming 6G's security and privacy issues. Multi-layered intrusion detection and prevention employing deep reinforcement learning and Deep Neural Networks (DNN) can protect 6G against IP spoofing attacks, flow table overloading attacks, Distributed Denial of Service (DDoS) attacks, control plane saturation attacks, and host location hijacking attacks. Machine learning (ML) techniques like Decision Trees and Random Forest can be also beneficial for detecting DDoS attacks in SDN systems because of their speed and accuracy.

This chapter aims to provide a discussion of security and privacy issues in 6G and how the integration of blockchain and AI with 6G can provide effective solutions to overcome these issues. The chapter starts by providing an overview of wireless communications technologies from 0 to 6G. This is followed by discussing the main security and privacy issues in 6G networks. Then, the integration of blockchain with 6G will be discussed by highlighting possible solutions to overcome security and privacy issues associated with 6G. The integration of 6G with AI will also be discussed by highlighting the importance of AI in 6G and how AI with 6G can provide better and effective security and privacy solutions. In the end, healthcare with 6G is presented as a use case by highlighting security issues and discussing the role of AI and blockchain in providing effective security solutions in the healthcare sector.

The remainder of this chapter is organized as follows: Sect. 2 presents an overview of mobile communication generations from 0 to 6G; Sect. 3 discusses security and privacy issues of 6G; Sect. 4 presents the integration of blockchain with 6G; Sect. 5 discusses the integration of AI with 6G; Sect. 6 presents use case of healthcare with 6G, and Sect. 7 is the Conclusion.

## 2 From 0G to 6G: An Overview

The evolution of mobile communication technologies passed through different phases. This section provides an overview of these different phases by highlighting the main advantages and drawbacks introduced by each generation. The discussion started from 0G or pre-cellular mobile telephony in 1970 to 6G that expected to be implemented in 2030.

0G or Zero Generation refers to the pre-cellular mobile telephony technology used in the 1970s such as Radio telephones and the telephone in cars, which was introduced before the invention of cell phones. 0G utilized technologies including IMTS (Improved Mobile Telephone Service), OLT (Norwegian for Offentlig Landmobil Telefoni), PTT (Push to Talk), AMTS (Advanced Mobile Telephone System), and MTS (Mobile Telephone System) [8]. Early example that utilized 0G technology was

Autoradiopuhelin (ARP), which was introduced by Finland in 1971 and become the first public commercial mobile phone network. Also, Germany introduced B-NetZ in 1972 as a public commercial mobile phone network [9].

1G or First-Generation technology for cellular networks was presented in the 1980s. Nippon Telephone & Telegraph (NTT) in Tokyo, Japan, introduced the world's first cellular system in 1979. Also, Nordic Mobile Telephone (NMT) and (TACS) in 1980 started to introduce their cellular systems across Europe [8]. 1G was mainly designed for voice services in which it uses analogue signals to transmit information [10]. 1G was the first step in the road to provide public and commercial mobile telephony. Although all 1G systems provided handover and roaming capability, the interoperability between countries was the major issue that faced 1G. This allows users to make voice calls only in one country. Also, 1G suffered from the lack of security and privacy as voice calls were transmitted back to the radio towers, which make these calls susceptible to various attacks. Also, since voice calls are not encrypted, the data transmission and calls can neither be secure, nor private. 1G also suffered from poor voice links and low capacity and unreliable handoff [11].

2G or Second-Generation technology was introduced in the early 1990s. The first 2G system was introduced by Finland in 1991. 2G is mainly based on digital modulation techniques, which enable both voice and short message services [12]. GSM (Global System for Mobile Communications) is the most well-known and commonly utilized 2G mobile communication system [13]. GSM overcomes the limitation of 1G by allowing international roaming between phone providers, which enable users to utilize their phones in different places around the globe. Compared to 1G, 2G provides better security and privacy since the encryption is applied on all text messages, which provide better security and allow only interned receiver to obtain messages and read it. Although 2G provides better security over 1G, it still has some issues such as crypto flaws, eavesdropping attack, SIM attack, fake base station (BS), absence of replay protection and DOS attack.

3G or Third-Generation technology of mobile communication was first introduced in 2000. It was the first wireless communication technology that enables data transfer of 2 Mbps [14]. 3G provides several advantages over 2G technology in which it provides faster communications that enable sending and receiving a large amount of data and enable high-speed video conferencing and 3D games. Also, with increased bandwidth and broadband capabilities with 3G, this allows web-based applications and audio and video streaming to work efficiently. On the other hand, 3G presents some challenges or drawbacks. One of the major issues is energy efficiency in which 3G consumes significantly more power than 2G, so 3G needs different devices and drivers. This makes building an infrastructure for 3G more challenging. Also, 3G needs high bandwidth requirements and costly fees for 3G licenses and agreements [15]. UMTS (universal mobile telecommunications service) or what was called W-CDMA is the primacy technology in 3G. It represents the recessive that is compatible with prior generations of wireless technologies through its heterogeneity with the legacy GSM and AMPS technologies. The evolution of UMTS into high-speed packet access (HSPA) and advanced HSPA (HSPA + ) allowed for greater end-to-end network efficiency and eventually led to the advancement of the next generation of networks [16].

4G or Fourth-Generation technology of mobile communication was first introduced in Stockholm in 2009. 4G represents the communication standard that delivers demands for broadband data transmission and broadcasting. 4G was the main driver for the Internet of Things (IoT) that enables users and physical objects to connect anytime, anywhere using any network path. 4G is the successor of 3G that is produced to overcome limitations associated with 3G and enable broader bandwidth, better security and privacy and high-speed internet access. According to the ITU union, 4G provides a data rate of 100 Mbps [17]. 4G is mainly based on the invention of LTE (Long-Term Evolution), which provides an IP-based invention for data transmission. LTE provides seamless mobility, QoS, and low latency for packet-switched traffic. 4G provides high-speed data rates up to 1Gbps, which provide high quality for audio and video streaming applications and online games. 4G also provides better security and privacy. On the other hand, 4G introduces some issues in which 4G consumes more power and need complicated and expensive hardware for implementation [8].

5G or Fifth-Generation technology of mobile communication was first introduced in the late 2010s. The demands for considerably faster data rates for users, and the exponentially growing IoT applications and services have meant the telecom industry has had to evolve around these pressures. 5G can satisfy the demands of the public and companies by creating a truly mobile and wired connected community. This will only be achieved by improving the radio access interface, which requires a new larger range of frequency, expanded bandwidth for the increasing user base, and support for the ever-expanding IoT [18]. There are various advantages to 5G. It transports a world with much improved mobile data broadband, ultra-responsiveness, ultra-reliability, ultra-low latency, ultra-fast data rates, and huge IoT capabilities. 5G can support around one million per square kilometre while 4G can only support about 4,000 devices. This can enable more video and audio streaming without disruption. Massive MIMO (Multiple Input Multiple Output) is a new digital technique used in 5G that uses multiple targeted beams to highlight and follow users throughout a cell site. Coverage, speed, and capacity are all improved as a result [19]. 5G is extremely compatible with WWWW (Wireless World Wide Web), which will create several applications using the high capabilities it provides.

6G or Sixth-Generation technology of mobile communication attracted the attention of several researchers, although 5G coverage is not yet being provided completely. The research has been started in 2019 to develop the 6G wireless technology. 6G provides new attractive characteristics that will integrate capabilities of previous mobile communication technologies such as high reliability, massive connectivity, high throughput and network densification. 6G with new capabilities will be able to provide several benefits to various applications such as AI, smart wearables, implants, autonomous vehicles, computing reality devices, sensing, and 3D mapping [3, 4]. 6G will improve performance and maximize user QoS. 6G is predicted to offer 1000 times faster wireless connectivity than 5G. Additionally, 6G is projected to enable ultra-long-range communication with a latency of less than

1 ms [5]. The average experienced downlink (DL) data rates for 6G are expected to be 100 Gbps and the number of devices per km$^2$ will be $10^9$.

The development of mobile communication technologies passed through various stages with each generation provided added capabilities, as shown in Fig. 1. The evolution started with supporting only voice calls with 1G and over the time the next generation focused on increasing bandwidth and speed of data transmission that allowed the appearance of novel web and internet applications. The developments continued to generate 5G that is the main driver for massive broadband and various IoT applications and services. The expectation talks about 6G and how it can change how businesses work by allowing automation and AI and creating a connected world.

Table 1 provides a summary of the advantages and disadvantages of each mobile communication technology that was previously introduced. Besides, a comparison of various features of 4G, 5G and 6G is presented in Table 2.

## 3 Security and Privacy Issues in 6G

Although 6G provide countless benefits by providing ubiquitous access and connectivity to various applications, security and privacy are still major issues that need to be addressed to provide more trust for the customers. This section discusses some of the security and privacy threats that are presented by 6G.

### 3.1 Security Issues in 6G

The old generation of mobile communication technologies (i.e., 1G, 2G, 3G) did not consider the security and privacy of users in their priorities. This makes these technologies suffer from severe security and privacy issues including cloning, eavesdropping, authentication and authorization issues. Because of the execution of wireless applications, 4G began to consider security, although it faced numerous security and privacy issues [20]. Then, 5G and 6G did the same but with security threats that need to be handled to take all the benefits of 6G wireless networks. Looking at the security in 6G should be combined with AI as 5G and 6G have almost fully integrated with AI, which led to what is called security automation. At the same time, adversaries with the power of AI have become more powerful and intelligent that makes detecting their attacks not an easy task.

Flash network traffic is one of the common vulnerabilities in both 5G and 6G networks. With the growing amount of IoT devices, the network capacity has to reach a demand fast. Wireless local area networks or even Femtocells, for example, are preferred when trying to increase the capacity of a network; However, transmitted information intended for the certified user equipment is more vulnerable to eavesdroppers and unauthorized users. This is because of broadcasting features of

**Table 1** Summary of advantages of different mobile communication technologies from 1 to 6G

| Technology | Advantages | Disadvantages |
|---|---|---|
| 1G | The first technology to provide public and commercial mobile telephony<br>Allows voice calls<br>Use analogue signal | Unable to interoperate between countries<br>Low spectral efficiency<br>Major security and privacy issues<br>Limited capacity<br>Large phone size<br>Poor handoff reliability |
| 2G | Use digital signals<br>Allows services including text messaging, image messages, and MMS<br>Provides better quality and capacity<br>Better response time. 10 times better than 3G | Unable to support internet and e-mail services<br>Limited data rates<br>Security and privacy issues<br>Unable to handle complex data such as videos<br>Require strong digital signals |
| 3G | Provide high-speed network communication for data transmission<br>Faster than previous networks<br>Provide very good voice call and huge MMS | Failure of WAP for internet access<br>Consumes more power<br>Require installing new 3G equipment<br>Require different handsets<br>Require closer base stations |
| 4G | Very high voice quality<br>High bandwidth<br>10 times faster than 3G<br>Support interactive multimedia and other broadband services<br>Easily access the Internet and other services | Require new complicated and expensive hardware as it uses new frequencies<br>Consumes more power<br>High cost for users<br>Previous components are not compatible with 4G |
| 5G | Low latency capabilities<br>Increased speed and bandwidth<br>Increased connectivity and access to various online services<br>Provides energy efficiency plans<br>Improved WAN connection<br>Data bandwidth of > 1 Gbps<br>Dynamic information access | Limited coverage as it is not yet implemented fully<br>Security and privacy issues<br>Higher expensive and complicated hardware equipment<br>Not compatible with previous technologies, so need new equipment |
| 6G | Provide ultra-broadband internet services<br>Very low latency capabilities<br>Support higher number of users per $km^2$<br>Higher downlink data rate that is expected to be 100Gbps<br>Uses visible lights which leverage the benefits of LEDs<br>Uses THz (Terahertz) frequencies, which has several benefits<br>Support home automation and other IoT applications | Need expensive and complicated hardware equipment<br>Security and privacy issues<br>Since it uses THz frequencies, it will face issues of THz, which is more dangerous for the environment<br>Lack of information control<br>Since it uses light, it will face issues of VLC<br>Require new expensive equipment |

**Table 2** Comparison of 4G, 5G and 6G

| Feature | 4G | 5G | 6G |
| --- | --- | --- | --- |
| Development period | 2009 | 2018 | 2030 |
| End-to-end (E2E) latency | 100 ms | 10 ms | 1 ms |
| Mobility support | Up to 350 km/hr | Up to 500 km/hr | Up to 1000 km/hr |
| Satellite integration | No | No | Fully |
| Per device peak data rate | 1 Gbps | 10 Gbps | 1 Tbps |
| XR | No | Partial | Fully |
| THz Communication | No | Limited | Widely |
| Haptic Communication | No | Partial | Fully |
| Maximum spectral efficiency | 15 bps/Hz | 30 bps/Hz | 100 bps/Hz |
| Architecture | MIMO | Massive MIMO | Intelligent surface |
| Maximum frequency | 6 GHz | 90 GHz | 10 THz |
| AI | No | Partial | Fully |
| Autonomous vehicle | No | Partial | Fully |
| Service level | Video | VR, AR | Tactile |
| Applications | High-speed applications, mobile TV, wearables | High-resolution video streaming, remote control of cars, robots and medical procedures | AI, smart wearables, implants, autonomous vehicles, computing reality devices, sensing, and 3D mapping |

wireless communications and open system architecture. The traffic from servers to machines must be protected to prevent traffic sniffing attacks [21].

DoS and DDoS are other common security threats in 6G. It is one of the most obvious attacks that occur to any device connected to the network and can damage the system as a whole. The use of various radio access network technologies is another security threat. Different radio access network technologies are starting to be integrated with 6G, these include mmWave, Wi-Fi, NB-IoT and Li-Fi. There are potential risks with the integration of these technologies and could pose a risk to the 5G network when systems are not tested or trailed to see whether or not they have any unknown/known vulnerabilities [22].

Integrity-based attack is one of the common attacks in 6G networks. This type of attack is implemented using a fake eNodeB, when consumer equipment is convinced to connect to the network, this is known as a malicious relay. This malicious relay is using the network as a Man-in-the-Middle attack. Because there is no integrity defence on this channel the attacker can take advantage of this, the intruder, who has

**Table 3** Summary of some of security attacks/threats in 6G

| Security threat in 6G | Description |
|---|---|
| Flash network traffic | Transmitted information needed for the intended user is vulnerable to eavesdroppers and unauthorized users. This is because of broadcasting features of wireless communications and open system architecture |
| Poisonous attack in AI | This attack is performed by tampering with malicious samples in the training data, which affect the learning outcomes of the AI system, which result in misclassification or incorrect regression outcomes |
| DoS and DDoS | DoS and DDoS attacks are carried out by continuously injecting fake heavy load in virtual network functions (VNFs) |
| Evasion attack in AI | This attack is similar to the poisonous attack but with tampering with the testing data instead to affect the result of the AI model |
| AI/ML frameworks | Most AI models use existing or traditional AI/Machine Learning (ML) frameworks that have several vulnerabilities that can be used as a way to hack or attack the integrity of data |
| 51% attack | Controlling the public blockchain with at least 51% of its mining power allows the attacker to gain control over the blockchain and manipulate its blocks and transactions |
| Quantum collision | In a quantum environment, a quantum collision occurs when two entirely independent inputs of a hash function produce the same result/output |
| Access control | Access control policies are broken, data or user credentials are stolen, and unauthorized resources or system parameters are accessed or modified by adversaries |
| Eavesdropping | Although transmissions in narrow beams with strong directionality are resistant to interception attacks, rogue nodes can still intercept the signal |

access to the intended user's encrypted communications, manipulates or changes the transmitted information such that the attacker fabricates the message that enters the intended receiver [23]. Other security attacks in 6G are summarized in Table 3.

## 3.2 Privacy Issues in 6G

6G is expected to provide ubiquitous access to various devices, objects, sensors, and autonomous applications. This will enable various devices in different IoT applications to share a huge quantity of data between various applications. However, this type of data can be personal data including financial data, habits, etc. [24]. For instance, although a smart light, which turns on and off as you move, provides an effective way of utilizing the energy, it can be used to identify the pattern of life in your house to identify which room you use and when and if there are other people in the house. Although the energy service provider can utilize your data to provide more efficient lighting service, for example, and create more customized and individually

**Table 4** How blockchain can address the challenges of 6G

| 6G Issues | How blockchain can address the challenge |
|---|---|
| Availability and transparency | The data in the blockchain network do not rely on centralized authority. Instead, the blockchain stores and distributes data across the network, where all the information are accessible across the blockchain network [42]. Additionally, smart contracts in the blockchain can provide reliable data exchange between the users [36] |
| Data integrity | The data in the blockchain have a unique encrypted hash to guarantee its integrity [43] |
| Access control | Smart contracts are capable of providing direct access between requestors and data centres with no need for a central authority [36] |
| Authentication | The blockchain-based public-key infrastructure is capable to identify entities in the network [7] |
| Privacy | The blockchain is capable of providing better and effective privacy in many forms. Firstly, the data are stored in a time-stamped and immutable mode to prevent data modification. In terms of data exchange across the network, smart contracts can protect data from malicious users by providing user authenticity [44]. Also, blockchain could enforce user data privacy policies using smart contracts by tracking and log the transaction across the network [42] |
| Scalability | The blockchain improves scalability through its ability to share the services between edge nodes across the network, interoperability across devices, and the link between IoT devices with no need for trusted intermediaries [38] |
| Security | The blockchain is a decentralized and distributed ledger in which transactions are added to the network and confirmed by a majority of the nodes participating in the network. SHA-256 hash was then used to link the new block to the preceding one. As a result, the blockchain provides a secure and immutable environment [36] |

personalized services based on personal preference, privacy is still the nightmare to customers and how service providers can leak their data and identifying the amount and type of information to be collected by these devices or services [25].

With the increase of adoption of AI-enabled smart applications that need situational and context-aware services, the traditional privacy-preserving techniques will not provide the required functionality due to various privacy issues encountered with 6G. Therefore, there is a need to adopt new technologies that provide better security and privacy. One of these technologies is Distributed Ledger Technology (DLT) such as blockchain, which provides more effective security solutions and provides privacy-preserving data sharing using DLT security features such as integrity, immutability, accountability and traceability [26, 27]. Other privacy-preserving techniques like Privacy protection using differential privacy (DP) also can provide an effective privacy-preserving solution for 6G wireless applications. Before sending the final output to the specified server, DP disturbs the actual data using fake design random noise functions. This eliminates the need for statistical analysis of the data collected and the inference of any personal information [28]. These approaches can be a solution to the privacy nightmare in 6G, but this does not change the fact that there is a need for new privacy-preserving technologies that can provide effective privacy solutions with the ubiquitous access and connectivity provided by 6G.

## 4 Blockchain for 6G

Adopting one of the technologies of DLT such as blockchain can provide better security and privacy for 6G. Blockchain provides effective security solutions utilizing immutability, tamper-proof, integrity, and accountability features. This section presents the integration of blockchain with 6G. It begins by outlining blockchain and its characteristics, benefits/advantages, integration with 6G and how this integration can utilize various features of blockchain in 6G.

### 4.1  Overview of Blockchain

Blockchain technology is relatively a recent technology that has been grasped by several scholars to explore more after success in the financial sector, cryptocurrency. In 2009, an anonymous person namely Satoshi Nakamoto published a white paper that solved a double-spending problem in an electronic cash transaction. It allows individuals to receive or send online payments without financial intuition using blockchain technology [29].

Blockchain is defined as distributed and decentralized ledger in which each node in the network contributes to managing transactions. The transactions are encrypted and stored in a block. Each block is time-stamped and encompasses a hash function of the preceding blocks to link the blocks together. The time-stamp and hash functions are used to protect the integrity and immutability of the transactions that are distributed across the network. To keep the network secure, the majority of the nodes in the network should agree to add a block to the chain through a mechanism called a consensus algorithm. The consensus verifies and authenticates the transactions without a central authority [30].

### 4.2  Benefits of Blockchain

The structure of building a blockchain network gives its own benefits that would not be found in other technologies. Although blockchain is one of the common

DLT technologies, it has its key benefits. This section provides the key benefits of blockchain including:

- **Persistency**: After validating and adding a block into the chain by the agreement of the majority of participant nodes in the blockchain network, data could not be modified or deleted because each block contains a timestamp of creating the block and hash function of all blocks in the chain. This maintains the integrity of the data [31].
- **Auditability**: The auditability in blockchain comes from the fact that each block stores the hash value of the previous block, which allows the current block to be connected with the previous block. This allows the data to be verified and tracked back [31].
- **Availability**: Because the data are shared among all nodes in the blockchain network, it can be accessed from other nodes even if one node is down.
- **Decentralized Management**: The automation of management comes from using a consensus protocol where each node in the network must follow the protocol to run the network. The consensus is an agreement between the nodes to validate the transaction and adding the block to the chain [32].
- **Transparency:** Because all data are shared throughout the network nodes, every participant node has access to it, blockchain provides a high level of transparency [33].
- **Privacy**: Privacy could be a great challenge in the blockchain because of the transparency factor [32]. However, with Public Key Infrastructure (PKI) and adding smart contracts in the blockchain network, privacy could be enhanced. PKI provides each user with their unique public key to represent their identity whereas, smart contracts could be used to add access to the data, enforce privacy policy, and add automation network tracking.
- **Security**: Blockchain provides a high level of security in two forms: PKI and cryptographic. Every user in the network has its own unique public key to represent the identity that protects them from malicious attacks. The cryptographic is used to encrypt data using SHA-256 that protects the data from being deleted or modified [32].

## 4.3   Integration of Blockchain with 6G

Despite the recent development of the 5G mobile network, academia and industry have commenced envisions in the next generation; the 6G mobile network by reviewing the development of the 5G [34], and highlighting the requirements in the 6G [22]. The requirements of 6G will enhance in terms of data rate, high system capacity, efficient high spectrum, lower data delay, and expected wider and deeper coverage. These requirements are a response to the emergence of new technologies, such as the Internet of Vehicles and the Internet of Everything that require high convergence of a massive number of sensors and devices, which are far beyond the capability of 5G [35].

The envisioned 6G mobile network would enable system services beyond 5G. However, the limitation in the technology adopted in 5G could not enable to reach expectation on 6G because of the extraordinary explosive growth in mobile traffic [36]. The growth has led to perceptible challenges in developing the next-generation mobile network. These challenges are not only related to the massive connectivity but are related to security and privacy [36]. The unprecedented in increasing mobile users, machine-to-machine and device-to-device connections, and the number of transactions, scalability, minimize latency, higher throughput of the network are main challenges in the future generation network [37]. These challenges rise series of security and privacy against cyber-attacks, protecting the massive data transmission between devices, and preventing unauthorized manipulation [38]. Therefore, integration with enabler technologies is essential to address the challenges.

5G provides network services through several underlying technologies, such as cloud computing. Although cloud computing paved the 5G to its objectives such as unlimited storage and communication power, the centralized architecture remains an obstacle to response to the massive amount of data and IoT devices [34, 36]. The challenges in centralized cloud computing remain unsolved in terms of security and privacy. Single point of failure and cyber activities lead to failure in security services such as data availability, privacy, and data integrity. Additionally, cloud services are provided through cloud service providers and users' data are stored in the cloud. These raise an issue in privacy services such as the method used to access the data, and the personal information of the users [36].

The integration with other enabler technologies such as decentralized architecture such as blockchain will bring the prospect to tackle the challenges. Blockchain is distributed, decentralized, and trustless network that operates without a centralized authority. This could improve transparency by sharing the transaction details between the participants in the network, provides rigorous security to prevent cyber-attacks such as DoS and privacy to protect sensitive information and prevent data manipulation [6].

Many studies have shown that blockchain could address the limitations associated with 6G [37–40]. Blockchain is capable of storing data across the network immutably and securely. This eliminates the single point of failure and data manipulation by malicious attacks, which ensures data availability and data integrity [7]. Additionally, blockchain is used to enhance the accessibility and authentication of data using smart contracts [27, 41]. Smart contracts can add restrictions to access data through decentralized rules.

Maintaining data transparency is a critical issue regarding data modification, user personal information, and access control, however, blockchain technology has the capability of maintaining data transparency while considering the security and privacy of data. This is due to many reasons. Firstly, every transaction on the blockchain network is approved through a consensus mechanism and agreed upon by the majority of the nodes in the network. These transactions are stored in a block and add blocks to the previous ones using SHA-256 encryption. Additionally, every user in the network has their public key to represent their identity and keep their personal information anonymity. Lastly, smart contracts are capable of providing restrictions to whom to access the data and can enforce privacy policies to access the data [36].

Moreover, integration of 6G with blockchain would enhance the network performance and increase scalability with low-latency services. The scalability improvement would be enhanced through the integration of the blockchain with mobile edge computing. Instead of centralized computing, a distributed structure would enable to share of reliable resources between edge nodes. Accordingly, the interoperability across devices could be improved because of the structure of the blockchain [38].

## 4.4 Blockchain for Better Security in 6G

Security issues are critical when it comes to mobile communication networks. It could be range from DoS, single point of failure, data tampering, and data leakage. Many studies illustrated that security issues still existed in the deployment of 5G [22, 36]. However, 6G is expected to work with trillions of devices and millions of mobile users. Investigation of these security issues and tackle them before developing the 6G is vital.

DoS attack is one of the challenging threats in 6G. Although many solutions have been proposed to mitigating the issue such as using AI techniques to detect the DoS, this issue is going to be crucial challenging in the 6G due to the massive connectivity. This threat could result in low latency to mobile uses, affect the scalability, and availability of the network [45].

The single point of failure threat is a result of the dependency on centralized systems. Even though the centralized systems in the computing model, edge computers, and others provide services that could not be available in previous networks such as on-demand and minimized the management efforts [36], the expansion of mobile uses and IoT devices would not be met. This threat could affect the performance and the security services in the network [39].

Data tampering threat concerns are about modifying the data by malicious attacks across the mobile network. Because the network services are remote communication between many actors, IoT devices, and intermediate provider services, providing data integrity is important. Although many cryptographic tools are used to protect the data integrity such as keyless signature infrastructure (KSI) or using a third party to validate data, these techniques would not be efficient with the massive increase in connectivity [7]. Data leakage threat is highly vulnerable in data sharing in the networks because of the cyber-attacks. This could lead to losing valuable customer personal information in the mobile networks, or location information of cars in vehicular networks. Data leakage is high-risk in centralized systems where data are stored in a single location. This model put the entire database at risk [46].

Blockchain is one of the promising technologies to tackle security threats. Blockchain features such as data sharing, the protocol to preserve the data, and access control are innovative features to enhance security. Firstly, instead of relying on a single point, blockchain is decentralized distributed ledgers across a network of nodes. This structure provides robustness and availability of data stored in the blockchain network. Additionally, the data in the network are immutable, which could not be changed or modified since each block is hashed with time-stamped and linked to the previous ones in such a way that could not modify the content of the blocks. Furthermore, the users in the network are using PKI to recognize their identities. Lastly, the network is not controlled by a single entity. Instead, the network is autonomous management using a consensus mechanism that provides a guarantee for agreement between participant nodes [47].

The integration of 6G with blockchain would help to tackle the limitations in network security. The decentralized and distributed structure would help to overcome the single point of failure and DoS. The immutability of data increases the integrity of the data and prevents it from being changed by malicious attacks [7].

## 5   AI for 6G

Because AI can learn to achieve self-configuration, self-optimization, self-organization, and self-healing, and therefore improve feasibility, intelligence is a vital element of 6G networks. This section discusses the integration of AI with 6G and how this integration can overcome security and privacy issues associated with 6G.

### 5.1   An Overview of AI

AI is an emerging and rapidly developing technology. Its applications are becoming more prevalent with every passing day. AI is an exciting emerging technology with many possible applications that could revolutionize society. It is predicted that the number of AI digital voice assistants will surpass Earth's human population by 2024 and hit 8.4 billion units [48]. AI can play games, aid healthcare, drive cars, assist law enforcement, and possibly control autonomous weapons. Considering its potentially powerful applications, AI, therefore, must be developed to ensure it is safely implemented. Some experts argue that confidentiality, integrity, and availability (CIA) are key security requirements for AI [49]. Ensuring the integrity of AI output is a clear issue and of huge importance, because AI can interfere with digital media [50].

AI is defined as "*computers that are able to perform tasks typically carried out by humans*" [50]. Due to the exponential growth of computing performance outlined by Moore's Law, some predict that computers with human general intelligence could exist within the next twenty years [51]. The concept of AI has been known for at least 80 years. The 1950 paper by Alan Turing, considered the possibility of real thinking machines. Turing devised a test for assessing a machine's ability to think. A benchmark was proposed that if a computer can fool an interrogator at least 30% of the time, then the test is passed.

A significant subset and evolution of AI is ML. Key developments were made in the 1950s–1970s [51]. ML can be described as a process of complicated computation involving intelligent decision-making and automated pattern recognition based on training sample data. It is considered a supervisory method because it requires initial data to be supplied by human users. ML plays a role in cybersecurity, for example, the use of Artificial Neural Networks (ANNs) in misuse/signature detection systems [52].

## 5.2  Integration of AI with 6G for Effective Security

Implementing 6G networks needs AI to empower autonomous networks. Hence, security attacks on AI systems particularly ML techniques will impact 6G. This includes poisoning attacks, logic corruption data injection, model inversion, model evasion, and membership inference attacks [53]. However, the integration of AI with 6G will provide several advantages to overcome security and privacy challenges in 6G. For example, IP spoofing, flow table overloading, DDoS, control plane saturation, and host location hijacking attacks can all be prevented by multi-layered intrusion detection and prevention employing deep reinforcement learning and Deep Neural Networks (DNN) [54]. Furthermore, because of their speed and precision, DDoS attacks in SDN systems can be detected using machine learning techniques such as Decision Trees (DT) and Random Forest (RF). ML-based adaptive security techniques are also effective against SDN/NFV threats, as 6G networks presume dynamic deployment of virtual services on-demand [55].

In contrast to current centralized cloud-based AI systems, 6G will rely heavily on edge intelligence. In the vast device and data regime, the distributed nature facilitates the implementation of edge-based federated learning to ensure communication efficiency [56]. 6G network architecture envisions connected intelligence and AI at multiple levels of the network structure. AI has the potential to prevent DoS attacks on cloud servers at the cellular level [57]. Also, in a mesh network, a device's multi-connectivity allows numerous base stations to utilize AI classification algorithms to analyse the device's behaviour and use weighted average ways to collectively decide on its authenticity [58].

Furthermore, utilizing AI-powered predictive analytics, attacks on the blockchain, such as 51% attacks, can be predicted before they happen. A quantum computer might jeopardize asymmetric key cryptography. They can, however, provide exponential speedups for AI/ML techniques, know how to complete previously impossible jobs much faster. As a consequence, quantum machine learning for network security can be a valid strategy against quantum computer-based attacks [59].

Also, in Visible Light Communication (VLC) systems, intelligent beamforming approaches based on RL give the best beamforming plan to protect from eavesdropper threats. Also, a possible option for detecting jamming attacks is anomaly-based detection systems using AI. Node compromise attacks can also be prevented with AI-based authentication and authorization systems [60].

AI also can provide better solutions for privacy issues in 6G. For instance, edge-based ML approaches can be utilized to detect privacy-preserving routes dynamically and transmit data over the highest rank privacy-preserving routes. Also, federated learning, as compared to cloud-based centralized learning, maintains data close to the user, enhancing data privacy and location privacy. Furthermore, due to the vast variety of applications in 6G and the massive data collection required to feed intelligent models, customers will demand varying levels of privacy on different applications. AI-based service-oriented privacy-preserving policy changes could enable fully automated 6G networks with retained privacy [61].

# 6   Use Case: Healthcare with 6G

Healthcare is one of the major applications that 6G can provide numerous advantages. 6G communication technology is projected to dramatically transform healthcare, with healthcare being wholly reliant on communication technology [62]. Telesurgery will be performed more efficiently thanks to high-speed communication provided by 6G. Also, 6G will provide numerous benefits in health monitoring with smart wearable devices [63]. 6G will also play a vital role in the development of the Intelligent Internet of Medical Things (IIoMT), which are mainly AI-driven intelligent machines that make their own decisions using communication technologies. 6G can also be utilized in the development of precision medicine, which can allow more precise and customized healthcare [64].

## 6.1   Security Issues in Healthcare with 6G

Although 6G provide numerous benefits in the healthcare domain, security is still one of the major issues that need to be addressed. Some security threats in healthcare include:

- **Data Integrity**: Protecting the integrity of patients' data is the priority for any healthcare provider. However, due to the high accessibility and connectivity provided by 6G, maintaining data integrity will be one of the main challenges in the healthcare system.
- **Phishing**: It is a popular method of stealing personal information, particularly employee information from healthcare firms. The phishing approach is a social engineering technique that manipulates people into sharing personal information. For example, a hacker may create a website that imitates an official website to obtain access to employee information and thus patient information [65].
- **DoS and DDoS**: These attacks are used to block authorized users from accessing the healthcare system by overloading the healthcare system with a flood of fake

traffic that shut down all healthcare services. DoS and DDoS will be serious attacks that target healthcare systems with 6G.

- **Social Engineering**: Humans are always the weakest point in the security chain. By using various social engineering attacks, hackers can deceive employees including doctors, nurses, etc. to steal sensitive information thanks to their lack of security awareness.
- **Eavesdropping**: With the use of 6G in healthcare, eavesdropping will be one of the security issues that result from signal interception due to the nature of wireless communication.
- **AI-based Attacks**: Although 6G utilize AI to provide several automation services in the healthcare sector, this can create several security vulnerabilities that can be used as a way to hack or attack the integrity of data.
- **Blockchain and AI for Effective Security in Healthcare with 6G**

Integrating blockchain with 6G in the healthcare system can provide effective security solutions to overcome most security challenges in the healthcare sector. Blockchain can overcome security issues associated with data integrity by gathering patient data and storing them in an internal catalogue. Then, smart contracts can be utilized to access patient data controlled by patients and their suitable policies. As a result, a patient can share his or her medical information on his or her own terms [66]. Patients can also have access control thanks to blockchain cryptographic keys. Each patient has a "master" key that allows them to "unlock" their health data and share a copy with healthcare providers as needed. Patients can limit their actions to reading or writing information, and they can revoke keys if the device on which the key is stored becomes hacked [67].

Also, blockchain stores data in a time-stamped and immutable mode to prevent data modification. Furthermore, with the help of AI with 6G, DoS and DDoS attacks can be detected and provide availability and accessibility of healthcare services 24/7 for authorized users. As blockchain is a distributed and decentralized ledger, a single point of failure associated with centralized systems will not exist and data can be accessed from any node in the blockchain network. The combination of 6G with AI can also provide an effective security solution to social engineering and phishing attacks by utilizing multi-layered intrusion detection and prevention techniques [68].

# 7 Conclusion

Although 6G provides unlimited advantages for various applications due to the huge capabilities it provides, security and privacy are the major issues that 6G need to address. Blockchain technology can play an important role to resolve some of these issues. The integration of 6G with blockchain can help to tackle the limitations in network security. The decentralized and distributed features can help to overcome the single point of failure and DoS attacks. The immutability of data can also increase data integrity. In the same regard, intelligence is a key feature of 6G networks, as

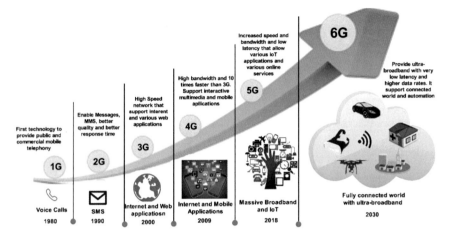

**Fig. 1** Development of wireless communication technologies from 1 to 6G

the integration of AI with 6G networks can learn to achieve self-configuration, self-optimization, self-organization, and self-healing, and ultimately improving feasibility. The integration can also provide several advantages to overcome security and privacy challenges in 6G. For instance, multi-layered intrusion detection and prevention employing deep reinforcement learning and DNN can protect 6G against DDoS, flow table overloading, control plane saturation, IP spoofing, and host location hijacking attacks. The chapter provided an overview of 6G by discussing the main security and privacy issues in 6G networks. Then, the integration of blockchain with 6G was discussed by highlighting possible solutions to overcome security and privacy issues associated with 6G. The integration of 6G with AI was presented by highlighting the importance of AI in 6G and how AI can provide better and effective security and privacy solutions in 6G. In the end, healthcare with 6G was presented as a use case by highlighting security issues and discussing the role of AI and blockchain in providing effective security solutions in the healthcare sector.

# References

1. Yang P, Xiao Y, Xiao M, Li S (2019) 6G Wireless Communications: Vision and Potential Techniques. IEEE Netw 33(4):70–75. https://doi.org/10.1109/MNET.2019.1800418
2. Letaief KB, Chen W, Shi Y, Zhang J, Zhang A (2019) The Roadmap to 6G: AI Empowered Wireless Networks. IEEE Commun Mag 57(8):84–90. https://doi.org/10.1109/MCOM.2019.1900271
3. David K, Berndt H (2018) 6G Vision and requirements: Is there any need for beyond 5g? IEEE Veh Technol Mag 13(3):72–80. https://doi.org/10.1109/MVT.2018.2848498
4. Elmeadawy S, Shubair RM (2019) 6G wireless communications: future technologies and research challenges. In: 2019 international conference electrical Comput. Technol. Appl. https://doi.org/10.1109/ICECTA48151.2019.8959607.

5. F. Clazzer, A. Munari, G. Liva, F. Lazaro, C. Stefanovic, P. Popovski (2019) From 5G to 6G: Has the Time for Modern Random Access Come?, Accessed: Aug. 10, 2021. https://arxiv.org/abs/1903.03063v1.
6. H. F. Atlam, A. Alenezi, M. O. Alassafi, G. B. Wills (2018) Blockchain with Internet of Things: Benefits, challenges, and future directions, Int. J. Intell. Syst. Appl., vol. 10, no. 6, doi: https://doi.org/10.5815/ijisa.2018.06.05.
7. Salman T, Zolanvari M, Erbad A, Jain R, Samaka M (2019) Security services using blockchains: A state of the art survey. IEEE Commun. Surv. Tutorials 21(1):858–880. https://doi.org/10.1109/COMST.2018.2863956
8. Meraj M, Mir I, Kumar S (2015) Evolution of Mobile Wireless Technology from 0G to 5G. Int. J. Comput. Sci. Inf. Technol. 6(3):2545–2551
9. Lopa M, Vora J (2015) Evolution of mobile generation technology: 1G to 5G. Int. J. Mod. Trends Eng. Res. 02:281–291
10. Wang M, Zhu T, Zhang T, Zhang J, Yu S, Zhou W (2020) Security and privacy in 6G networks: New areas and new challenges. Digit. Commun. Networks 6(3):281–291. https://doi.org/10.1016/j.dcan.2020.07.003
11. J. H. Schiller (2003) Mobile Communications. 2nd Edition, Pearson Education Limited.
12. Stojmenovic I (2002) Handbook of Wireless Networks and Mobile Computing. John Wiley & Sons Inc.
13. J. G. Shinde, P. Shamuvel, K. Sunil (2018) Review Paper on Development of Rice Transplanter, IRE Journals 2(5), pp. 94–100
14. A. K. Pachauri, O. Singh (2021) G Technology—Redefining wireless Communication in upcoming years, Int. J. Comput. Sci. Manag. Res., 1(1)
15. Nguyen VG, Brunstrom A, Grinnemo KJ, Taheri J (2017) SDN/NFV-Based Mobile Packet Core Network Architectures: A Survey. IEEE Commun. Surv. Tutorials 19(3):1567–1602
16. Gupta A, Jha RK (2015) A Survey of 5G Network: Architecture and Emerging Technologies. IEEE Access 3:1206–1232. https://doi.org/10.1109/ACCESS.2015.2461602
17. Nitesh GS, Kakkar A (2016) Generations of Mobile Communication. Int. J. Adv. Res. Comput. Sci. Softw. Eng. 6(3):320–324
18. Karla B, C. DK, (2014) A comparative study of mobile wireless communication network: 1g to 5g. Int. J. Comput. Sci. Inf. Technol. Res. 2:430–433
19. Eluwole OT, Udoh N, Ojo M, Okoro C, Akinyoade AJ (2018) From 1G to 5G, what next? IAENG Int J Comput Sci 45(3):413–434
20. Porambage P, Gur G, Osorio DPM, Liyanage M, Gurtov A, Ylianttila M (2021) The Roadmap to 6G Security and Privacy, IEEE Open. J. Commun. Soc. 2:1094–1122. https://doi.org/10.1109/OJCOMS.2021.3078081
21. Wang HM, Zheng TX, Yuan J, Towsley D, Lee MH (2016) Physical Layer Security in Heterogeneous Cellular Networks. IEEE Trans Commun 64(3):1204–1219. https://doi.org/10.1109/TCOMM.2016.2519402
22. De Alwis C (2021) Survey on 6G Frontiers: Trends, Applications, Requirements, Technologies and Future Research, IEEE Open. J. Commun. Soc. 2:836–886. https://doi.org/10.1109/OJCOMS.2021.3071496
23. V. Ziegler, P. Schneider, H. Viswanathan, M. Montag, S. Kanugovi, A. Rezaki (2020) Security and trust in the 6G era, NOKIA Bell Labs.
24. Ahmad I, Shahabuddin S, Kumar T, Okwuibe J, Gurtov A, Ylianttila M (2019) Security for 5G and beyond. IEEE Commun. Surv. Tutorials 21(4):3682–3722. https://doi.org/10.1109/COMST.2019.2916180
25. Y. Section, A. Gurtov, L. Mucchi, I. Oppermann (2020) 6G-White-Paper-Trust-Security-Privacy.
26. Atlam HF, Walters RJ, Wills GB (2018) Fog Computing and the Internet of Things: A Review, big data Cognitive. Computing 2(2):1–18
27. H. F. Atlam, G. B. Wills (2018) Technical aspects of blockchain and IoT, In Role of Blockchain Technology in IoT Applications, Advances in Computers, pp. 1–35.

28. Wang Q, Chen D, Zhang N, Ding Z, Qin Z (2017) PCP: A Privacy-Preserving Content-Based Publish-Subscribe Scheme with Differential Privacy in Fog Computing. IEEE Access 5:17962–17974. https://doi.org/10.1109/ACCESS.2017.2748956
29. S. Nakamoto (2009) Bitcoin: A Peer-to-Peer Electronic Cash System, DOI: https://doi.org/10.1007/s10838-008-9062-0.
30. Atlam HF, Wills GB (2019) Intersections between IoT and distributed ledger, In Role of Blockchain Technology in IoT Applications. Adv Comput 115:2019
31. Zheng Z, Xie S, Dai H, Chen X, Wang H (2017) An Overview of Blockchain Technology: Architecture, Consensus, and Future Trends, Proc. - 2017 IEEE 6th Int. Congr. Big Data, BigData Congr. 2017:557–564. https://doi.org/10.1109/BIGDATACONGRESS.2017.85
32. Tt K, He K, O.-M. L, (2017) Blockchain distributed ledger technologies for biomedical and health care applications. J Am Med Inform Assoc 24(6):1211–1220. https://doi.org/10.1093/JAMIA/OCX068
33. H. F. Atlam, M. O. Alassafi, A.Alenezi, R.Walters, G. B. Wills (2018) XACML for Building Access Control Policies in Internet of Things, In Proceedings of the 3rd International Conference on Internet of Things, Big Data and Security (IoTBDS 2018), pp. 253 -260.
34. Chen S, Liang YC, Sun S, Kang S, Cheng W, Peng M (2020) Vision, Requirements, and Technology Trend of 6G: How to Tackle the Challenges of System Coverage, Capacity, User Data-Rate and Movement Speed. IEEE Wirel Commun 27(2):218–228
35. Zong B, Fan C, Wang X, Duan X, Wang B, Wang J (2019) 6G Technologies: Key Drivers, Core Requirements, System Architectures, and Enabling Technologies. IEEE Veh Technol Mag 14(3):18–27. https://doi.org/10.1109/MVT.2019.2921398
36. Nguyen DC, Pathirana PN, Ding M, Seneviratne A (2020) Blockchain for 5G and beyond networks: A state of the art survey. J Netw Comput Appl 166:102693. https://doi.org/10.1016/J.JNCA.2020.102693
37. T. Hewa, G. Gur, A. Kalla, M. Ylianttila, A. Bracken, M. Liyanage (2020) The role of blockchain in 6G: Challenges, opportunities and research directions, 2nd 6G Wirel. Summit 2020 Gain Edge 6G Era, 6G SUMMIT 2020, doi: https://doi.org/10.1109/6GSUMMIT49458.2020.9083784.
38. Chowdhury MZ, Shahjalal M, Ahmed S, Jang YM (2020) 6G Wireless Communication Systems: Applications, Requirements, Technologies, Challenges, and Research Directions, IEEE Open. J. Commun. Soc. 1:957–975
39. Jiang W, Han B, Habibi MA, Schotten HD (2021) The Road Towards 6G: A Comprehensive Survey, IEEE Open. J. Commun. Soc. 2:334–366
40. Xu H, Klaine PV, Onireti O, Cao B, Imran M, Zhang L (2020) Blockchain-enabled resource management and sharing for 6G communications. Digit. Commun. Networks 6(3):261–269. https://doi.org/10.1016/J.DCAN.2020.06.002
41. H. F. Atlam, A. Alenezi, R. J. Walters, G. B. Wills (2017) An overview of risk estimation techniques in risk-based access control for the internet of things, In Proceedings of the 2nd International Conference on Internet of Things, Big Data and Security (IoTBDS 2017), pages 254–260.
42. A. Banerjee, K. P. Joshi (2017) Link before you share: Managing privacy policies through blockchain, Proc. - 2017 IEEE Int. Conf. Big Data, Big Data, pp. 4438–4447, doi: https://doi.org/10.1109/BIGDATA.2017.8258482.
43. Maksymyuk T (2020) Blockchain-Empowered Framework for Decentralized Network Management in 6G. IEEE Commun Mag 58(9):86–92. https://doi.org/10.1109/MCOM.001.2000175
44. Butt TA, Iqbal R, Salah K, Aloqaily M, Jararweh Y (2019) Privacy Management in Social Internet of Vehicles: Review. Challenges and Blockchain Based Solutions, IEEE Access 7:79694–79713. https://doi.org/10.1109/ACCESS.2019.2922236
45. Amaizu GC, Nwakanma CI, Bhardwaj S, Lee JM, Kim DS (2021) Composite and efficient DDoS attack detection framework for B5G networks. Comput. Networks 188:107871

46. X. Liang, S. Shetty, D. Tosh, C. Kamhoua, K. Kwiat, L. Njilla (2017) ProvChain: A Blockchain-Based Data Provenance Architecture in Cloud Environment with Enhanced Privacy and Availability, Proc. - 2017 17th IEEE/ACM Int. Symp. Clust. Cloud Grid Comput, pp. 468–477, doi: https://doi.org/10.1109/CCGRID.2017.8.

47. M. W. Akhtar, S. A. Hassan, R. Ghaffar, H. Jung, S. Garg, M. S. Hossain, (2020) The shift to 6G communications: vision and requirements, Human-centric Comput. Inf. Sci. 2020 101, vol. 10, no. 1, pp. 1–27, doi: https://doi.org/10.1186/S13673-020-00258-2.

48. D. Silver (2017) Mastering the game of Go without human knowledge, Nat. 2017 5507676, vol. 550, no. 7676, pp. 354–359, doi: https://doi.org/10.1038/nature24270.

49. Bhatnagar S (2018) The Malicious Use of Artificial Intelligence: Forecasting, Prevention, and Mitigation Authors are listed in order of contribution Design Direction. Cornell Univ, New York

50. Reedy P (2020) Interpol review of digital evidence 2016–2019, Forensic Sci. Int. Synerg. 2:489–520. https://doi.org/10.1016/J.FSISYN.2020.01.015

51. R. Anyoha (2017) The History of Artificial Intelligence - Science in the News, Harvard University, 28 August 2017, 2018. https://sitn.hms.harvard.edu/flash/2017/history-artificial-intellige nce/ (accessed Oct. 09, 2021).

52. Dua S, Du X (2011) Data mining and machine learning in cybersecurity. Taylor & Francis

53. Benzaïd C, Taleb T (2020) AI for beyond 5G Networks: A Cyber-Security Defense or Offense Enabler? IEEE Netw 34(6):140–147. https://doi.org/10.1109/MNET.011.2000088

54. Abdulqadder IH, Zhou S, Zou D, Aziz IT, Akber SMA (2020) Multi-layered intrusion detection and prevention in the SDN/NFV enabled cloud of 5G networks using AI-based defence mechanisms. Comput. Networks 179:107364. https://doi.org/10.1016/J.COMNET.2020.107364

55. Santos R, Souza D, Santo W, Ribeiro A, Moreno E (2020) Machine learning algorithms to detect DDoS attacks in SDN. Concurr. Comput. Pract. Exp. 32(16):e5402. https://doi.org/10.1002/CPE.5402

56. Lu Y, Huang X, Zhang K, Maharjan S, Zhang Y (2021) Low-Latency Federated Learning and Blockchain for Edge Association in Digital Twin Empowered 6G Networks. IEEE Trans. Ind. Informatics 17(7):5098–5107. https://doi.org/10.1109/TII.2020.3017668

57. Ma C (2019) On Safeguarding Privacy and Security in the Framework of Federated Learning. IEEE Netw 34(4):242–248

58. H. F. Atlam, M. A. Azad, A. G. Alzahrani, G. Wills (2020) A review of blockchain in internet of things and Ai, Big Data Cogn. Comput., vol. 4, no. 4, doi: https://doi.org/10.3390/bdcc4040028.

59. Biamonte J, Wittek P, Pancotti N, Rebentrost P, Wiebe N, Lloyd S (2017) Quantum machine learning, Nat. 2017 5497671, vol. 549, no. 7671, pp. 195–202, doi: https://doi.org/10.1038/nature23474.

60. Xiao L, Sheng G, Liu S, Dai H, Peng M, Song J (2019) Deep reinforcement learning-enabled secure visible light communication against eavesdropping. IEEE Trans Commun 67(10):6994–7005. https://doi.org/10.1109/TCOMM.2019.2930247

61. M. Liyanage, J. Salo, A. Braeken, T. Kumar, S. Seneviratne, M. Ylianttila (2018) 5G Privacy: scenarios and solutions, IEEE 5G World Forum, 5GWF 2018 - Conf. Proc., pp. 197–203. https://doi.org/10.1109/5GWF.2018.8516981.

62. Nayak S, Patgiri R, Member S (2020) 6G Communication Technology: A Vision on Intelligent Healthcare. IEEE Internet of Things Journal.

63. Saad W, Bennis M, Chen M (2019) A vision of 6g wireless systems: Applications, trends, technologies, and open research problems. IEEE Netw 1–9.

64. Reddy B, Hassan U, Seymour C et al (2018) (2018) Point-of-care sensors for the management of sepsis. Nat Biomed Eng 29(2):640–648. https://doi.org/10.1038/s41551-018-0288-9

65. Mohammad RM, Thabtah F, McCluskey L (2015) Tutorial and critical analysis of phishing websites methods. Comput Sci Rev 17:1–24. https://doi.org/10.1016/J.COSREV.2015.04.001

66. Partala J, Nguyen TH, Pirttikangas S (2020) Non-interactive zero knowledge for blockchain: a survey. IEEE Access 8:227945–227961

67. Ji Y, Zhang J, Ma J et al (2018) BMPLS: Blockchain-Based Multi-level Privacy-Preserving Location Sharing Scheme for Telecare Medical Information Systems. J Med Syst 42. https://doi.org/10.1007/S10916-018-0998-2

68. Saha A, Amin R, Kunal S et al (2019) Review on "Blockchain technology based medical healthcare system with privacy issues. Secur Priv 2:e83. https://doi.org/10.1002/SPY2.83

# Security, Privacy Challenges and Solutions for Various Applications in Blockchain Distributed Ledger for Wireless-Based Communication Networks

Vivekanandan Manojkumar, V. N. Sastry, and U. Srinivasulu Reddy

**Abstract** Blockchain is a secure computing technology which could be used in a variety of applications like finance, supply chain management, land registry, health care, and education. To obtain services from various service providers, users or consumers register with the Registration Center Authority (RAC) via., wireless communications networks. However, RAC has flaws such as insider threat and Single Point of Failure (SPoF). To circumvent these problems, if user registration based on blockchain for any service is required through wireless communication networks and also blockchain networks has the Distributed Ledger (DL) for storing identity information about the users or consumers. In this chapter, we will discuss about the security and privacy issues that come with each of the applications mentioned above and also some of the key challenges in each applications such as data confidentiality, integrity, authentication, access control, and privacy. Public Key Infrastructure (PKI) is giving the solution for data confidentiality, integrity, and authentication. Access control is processed by the access control policies (to restrict the user permission) and the privacy is achieved through secure authentication.

**Keywords** Privacy · Security · Distributed ledger (DL) · Authentication · Access control · Health care · Banking · Supply chain management · 6G · 5G

V. Manojkumar
Department of Computer Science & Engineering, School of Engineering and Applied Sciences, SRM University-AP, Amaravati 522502, Andhra Pradesh, India
e-mail: vmanojk88@gmail.com

V. N. Sastry
Center for Mobile Banking (CMB), Institute for Development and Research in Banking Technology (IDRBT), Hyderabad, Telangana, India
e-mail: vnsastry@idrbt.ac.in

U. Srinivasulu Reddy (✉)
Machine Learning and Data Analytics Lab, Center of Excellence in Artificial Intelligence, Department of Computer Applications, National Institute of Technology Tiruchirappalli, Tiruchirappalli, Tamil Nadu, India
e-mail: usreddy@nitt.edu

© The Author(s), under exclusive license to Springer Nature Singapore Pte Ltd. 2022     117
M. Dutta Borah et al. (eds.), *AI and Blockchain Technology in 6G Wireless Network*, Blockchain Technologies, https://doi.org/10.1007/978-981-19-2868-0_6

# 1 Introduction

## 1.1 Blockchain

Satoshi Nakamoto [1] introduced blockchain technology for Bitcoin cryptocurrency and the transaction is executed based on peer-to-peer mode without a centralized authority. Blockchain is a secure DL, which has a list of transactions and the transactions records are maintained in the growing chains of a block. Each block is secured by cryptography techniques for providing integrity to the transactions records. The new transaction records are added in the global blockchain based on consensus mechanisms, the consensus mechanisms are performed by the minors (who have the higher capacity of computing power to perform the consensus mechanisms). Apart from that, each block has the entire hash of the block and each block has the hash of the preceding block. The consensus mechanisms are "Proof of Work (PoW)", "Proof of State (PoS)", "Byzantine Fault Tolerant (BFT)-based", "Sleepy", "Proof of Elapsed Time (PoET)", "Proof of Authority (PoA)", "Proof of Reputation (PoR)," etc., There are three types of blockchains, namely "private blockchain", "public blockchain", "consortium blockchain". Private blockchain: The read permission is open to all or read permission is restricted to specific network of nodes and write permission is performed by a single organization. Public blockchain: It is open to anyone to write, send, and receive. A public blockchain network allows the block added in to the blockchain based on the successful completion of the consensus mechanism. Consortium blockchain: The set of selected nodes only have write permission in the blockchain network and read operation is open to all [41].

## 1.2 Blockchain Architecture

The blockchain has five layers, i.e., physical, network, consensus, propagation, and application. The physical layer consists of blockchain nodes. The nodes are interconnected to form a blockchain. In the blockchain network, there are two types of nodes: Full nodes and light nodes. The full node consists of a full copy of DL. The full node is participating in the mining process. If the full node is completing a successful mining process, the miners get a reward. The light nodes are not required for higher hardware. They maintain only the last block of the blockchain network. The information goes from the full node to the light node. Network layer: Data delivery services are provided in this layer. All the nodes communicate to each other to process consensus mechanisms. There is no central authority to approve the transactions. In blockchain networks, the decentralized procedure is to approve the transactions. Consensus layer: This layer is responsible for incentives to the minor nodes, who complete the successful mining process. Propagation layer: It has the rules that follow how the blocks and messages are propagated through communication protocols in the network [2].

## 1.3  Distributed Ledger (DL) Technology

The blockchain networks has the DL. DL is a more powerful technology and nowa-days many use cases are used for identity management and machine-to-machine transactions. "Decentralization", "immutability", and "distributed" are the proper-ties of the DL. In decentralized, there is no single entity to maintain the data, therefore it is available at network nodes based on blockchain types. As a result, it safeguards against single points of failure as well as insider threats. Immutability: The data is recorded in the DL and can be verified at any moment during the auditing process. Distributed: All participating network nodes have access to DL, and it is visible to all network nodes based on blockchain type. Consistency, Availability, and Partition Tolerance (CAP) are the properties of distributed systems, and these characteristics are applicable to DL. The consistency property states that all computing nodes have the most recent version of DL. Availability refers to the total amount of data that can be accessed via., DL in the network at any time. Partition tolerance means that if one or more computational nodes fail, other nodes will continue to function, ensuring that there is no SPoF. As a result, blockchain-based services are required for a variety of applications [3].

## 1.4  Wireless Communication Networks

In the 1980s, the first generation (1G) of mobile communication networks was intro-duced, and also it is used for voice-over communication. They used analog technology for mobile communications. In the second generation (2G) of mobile communica-tions based on analog to digital conversion with supported additional features such as voice over communication and Short Message Service (SMS) was introduced. In the third generation (3G) of mobile communications, video calling facility, mobile TV, broadband services and Multimedia Message Services (MMS) were introduced. Fourth generation (4G) was introduced with improved broadband services, video streaming with high definition, voice over IP, and online gaming. 5G supports mobile broadband enhanced features with 10 Gbps data rates, latency with up to 1 ms, and network availability of 99.99% [43].

## 1.5  Limitation of 5G and Opportunities of Blockchain in 6G

5G technology has limitations such as reliability, data-rate, latency, processing, avail-ability, global coverage, connection density, and ground over spanning. These limi-tations are overcome in the 6G communication network [43]. The 6G technology has the features such as data rates of 1Tpbs, delay below 10 ms, availability of network 99.99%, $10^7$ devices/km$^2$ connection in the Internet of Everything (IoE) environment. Nowadays, blockchain is an emerging technology and it is used in many industries such as business and academics etc., Blockchain contains a shared ledger and it provides properties such as non-replicated transactions, confidentiality-based authentication, and decentralization. The blockchain maintains many users in the distributed environment for 6G networks. The smart contracts are used to execute (simulationusly on between users) the transactions in blockchain [44]. 6G is used in various applications such as Internet of Everything (IoE), smart grid, UAV, autonomous vehicle, Industry 5.0 and health care [43]. In each application, DL technology is used for various purposes as discussed in Sect. 4.

Now, we discuss the privacy and security challenges connected with 6G-based applications in this chapter, as well as solutions to those challenges. Some of the security and privacy key challenges in each application are "data confidentiality", "integrity", "authentication", "access control", and "privacy". Public Key Infrastructure (PKI) is giving the solution for data confidentiality, integrity, and authentication. Access control is processed by the access control [17] policies (to restrict the user permission for accessing the resources) and the privacy is achieved through secure authentication. We discuss the following major aspects in this chapter:

- The blockchain distributed ledger based 6G applications such as banking, finance, COVID-19, supply chain, land registration, health care, media and entertainment, and education sector;
- Each application's security and privacy challenges;
- Solutions for solving security challenges in blockchain-based applications.

## *1.6  Section Organization*

The following is a summary of the organization of the sections in this chapter. The mathematical preliminaries are presented in Sect. 2, and the system model is presented in Sect. 3. Blockchain Distributed Ledger (BLT) based 6G applications are shown in Sect. 4. Security analysis is presented in Sect. 5. The comparative research of blockchain-based security protocols is presented in Sect. 6. The future challenges of DL privacy and security are discussed in Sect. 7. Section 8 concludes with a conclusion.

## 2  Mathematical Preliminaries

The cryptographic techniques needed to solve security and privacy issues in blockchain [19] DL-based 6G applications.

## 2.1 Hash Function

**Definition 1** Hash function (H) takes an arbitrary length of the input such as a $\in$ $\{0,1\}^*$ and it produces a fixed size of the outputs such as b $\in$ $\{0, 1\}^n$ such that b = H (a). H has the following properties:

1. One-way property: For a given b, it is difficult to find the input message such that b = H(a).
2. Weak collision resistance: For a given input x, it is difficult to find input y i.e., $y \neq x$ such that $H(y) = H(x)$.
3. Strong collision resistance: It is difficult to find message1 and message2 pairs such that $H(message1) = H(message2)$.

The details about different hash functions are represented in Stallings [4].

## 2.2 Elliptic Curve Cryptography (ECC)

ECC is one of the public-key cryptography methods. It is based on the over finite field $(Z_P)$ using elliptic curves. The elliptic curve is defined as follows:

**Definition 2** The elliptic curve equation is $y^2 = x^3 + ax + b$ *over* $Z_P$, where p is prime and p > 3, x and y are solutions (x, y) $\in Z_P \times Z_P$ to the congruence $y^2 \equiv x^3 + ax + b \ (mod \ p)$, where a and b are constants a, b $\in Z_P$ such that $4a^2 + 27b^2 \neq 0$ (mod p), together with infinity in the point. A detailed explanation of the elliptic curve is given in Kobliz [5]. The elliptic curve is applicable for key exchange, encryption/decryption, and digital signature algorithms.

## 2.3 Bilinear Pairing

**Definition 3** Let $G_1$ and $G_2$ are additive cyclic groups over $E_p$ (a, b) and multiplicative cyclic groups over $Z_P$ with the order of large prime q. The bilinear map e: $G_1 \times G_1 \rightarrow G_2$ has the following properties:

Bilinearity: Let a, b $\in Z^*_p$ and A, B, C $\in G_1$. The bilinear paring (see Eq. 1):

$$e(\alpha + \beta, \gamma) = e(\alpha, \gamma) \cdot e(\beta, \gamma) \tag{1}$$

$$e(\alpha, \beta + \gamma) = e(\alpha, \beta) \cdot e(\alpha, \gamma)$$

$$e(a\alpha, b\beta) = e(a\alpha, b\beta)$$

$$= e(\alpha, \beta)^{a,b}.$$

Non-degeneracy: e $(\alpha, \alpha) \neq 1$.

Computability: e is to be computed efficiently. A detailed explanation is given in Menezes [6].

## 2.4  Fuzzy Extractor

**Definition 4** Fuzzy extractor is one of the techniques for user authentication using user biometric [7]. It has the generation function Gen(.) and reproduction function Rep(.), respectively. Gen(.) takes the biometric (BIO) from the user and it produces a biometric key, i.e., $\sigma_i \in \{0,1\}^a$ and public reproduction, i.e., $\tau_{i.}$. Rep(.) takes the user biometric (BIO') and public reproduction string $\tau_{i.}$ as an input and output is biometric key $\sigma_i$ such that the hamming distance is the calculated distance (BIO, BIO') < t, where t is a threshold value. A detailed explanation of the fuzzy extractor function is given in Dodis et al. [8].

## 2.5  Bio Hash Function

**Definition 5** Bio hashing is one of the techniques for user authentication using user biometrics [9]. It takes biometric input from the user and produces a particular code called biocode. The biocode is generated based on the user input biometric data with a random salt value. First, the feature is extracted from the fingerprint or face image. The biometric feature is added with the random salt value and finally the biocode is generated. A detailed explanation of the biohash function is given in Lumini and Nanni [10].

## 2.6  Chebyshev Chaotic Map

**Definition 6** Chebyshev polynomials (CP) is defined as $CP_n$ (x):[−1, 1]→[−1,1] based on degree n and it is represented as follows (See Eq. 2, 3, 4):

$$CP_n(x) = \begin{cases} \cos(n.arccos(x)) if\ x \in [-1, 1] \\ \cos(n\theta) if\ x = \cos\theta, \theta \in [0, \pi] \end{cases} \tag{2}$$

The CP can be defined as recursively as follows:

$$CP_n(x) = \begin{cases} 1 & if\ n = 1 \\ x & if\ n = 1 \\ 2x\,P_{n-1(x)} - P_{n-2(x)} & if\ n \geqslant 2 \end{cases} \tag{3}$$

The enhanced CP has the interval of $[-\infty, +\infty]$ and it is presented as follows [39, 40]:

$$CPa(CPb(x)) \equiv CPab(x) \equiv CPa(x)(CPn(x))(mod\ p) \tag{4}$$

p is a prime (large) number and $n \geq 2$, $x \in [-\infty, +\infty]$.

## 3   System Model with the 6G Applications

Figure 1 shows the system model. In the system model, the blockchain distributed ledger is applicable for 6G-based different applications such as finance (banking), e-voting, supply chain, health care, media and entertainment, land registry, education,

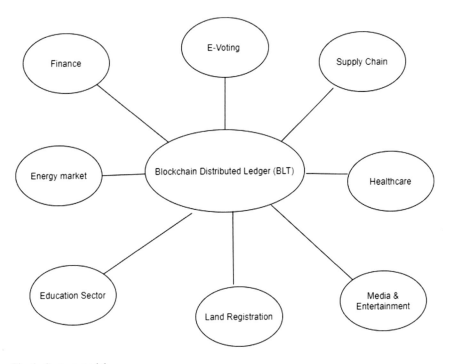

**Fig. 1** System model

energy management, and real estate. In each application, security and privacy are the major challenges. In Sect. 4, we discuss the privacy and security challenges of each application and we have given solutions for those challenges.

## 4   Blockchain DL Based 6G Applications

### 4.1   Blockchain DL in Banking

In banking, blockchain distributed ledger will play an important role. In general, all the bank transaction data is stored in a centralized server and all the customer data is also stored in a centralized server. Therefore, there is some possibility of SPoF and insider threats. In order to avoid these problems and to maintain all the customer data and customer transactions data, the decentralized server i.e., blockchain is needed. When compared to traditional banking and Fintech 1.0, the Fintech 2.0 (Blockchain + banks) provides a good customer experience, high efficiency, low cost, and distributed ledger safety [33].

### 4.2   Blockchain DL in Finance

In finance [32] industry, blockchain plays a vital role. Blockchain distributed ledger is used for many purposes such as identity management, integrity checking, machine to-machine communication, and multiparty computation. In identity management, storing the user data into a blockchain distributed ledger has security and privacy challenges. To solve the challenges of privacy and security, the user data is stored in the encryption format in a blockchain distributed ledger. In integrity checking, the part of user data is stored in the blockchain and at the time of verification the server will verify the integrity of user data. In machine-to-machine communication, the devices are registered with the blockchain in the IoT environment. At the time of authentication, devices are authenticate with each other without the help of GWN/RAC [35] in IoT. In multiparty computation, multiple nodes participate in the process of computation without trusting any authority. After successful computation by a minor node, that node will get incentives for the jobs [34].

### 4.3   Blockchain DL in COVID-19

Today's world is very much affected by the "Coronavirus (COVID-19)". Around 239,007,759 confirmed cases people were infected with the COVID-19 [38] and around 4,871,841 people died around the world due to coronavirus based on World

Health Organization (WHO) report reported on 14-10-2021 [37]. In this scenario, blockchain distributed ledger will play an important role. If any person is affected with the COVID-19, then their data (user locations, user information, etc.) stored in a blockchain distributed ledger will help to identify their location, contact tracing, home quarantine details, etc., For the user privacy, the user data encrypted is based on the cryptography techniques and health authority have the key of encrypted data. Whenever the health authority needs user data, data can have retrieved from blockchain DL and decrypt the data with the private key.

## 4.4  Blockchain DL in Supply Chain

DL is used in supply chain management to track product information such as where the product is created and delivered as well as the whole history of product information. Whenever product verification is needed, at the time the verification of product information is retrieved from DL. In food supply chain management [36], the blockchain distributed ledger maintains all the entities involved in the food supply chain. The entities are retailers, farmers, food manufacturers, and consumers. These entities are having the permission to access a full copy of the ledger. For others based on access write permission [42].

## 4.5  Blockchain DL in Land Registration

To avoid fraud and suspicious registrations, DL keeps a record of purchasing details (buyer and seller information) for the land registry.

## 4.6  Blockchain DL in Health Care

Patients' sensitive data is stored in DL in the healthcare sector; later, the data can be used for patient health analysis, and DL keeps track of the patient's data history, allowing clinicians to follow the exact health of the patient's condition. By using a private blockchain, different organizations will share the same distributed ledger for patient information. It has several security challenges such as data privacy and integrity. The patient's data is stored in blockchain based on the patient's permission. The patient's data is encrypted and it can be stored in a private blockchain to solve the privacy and security issues.

### 4.7   Blockchain DL in Entertainment and Media

If the entertainment media data is available in the blockchain DL, then the user have less latency and quality of media data accessing from the blockchain DL is increased. The blockchain distributed ledger is maintained in all minor nodes, so that the latency of media getting to the user has less latency when compared to centralized systems.

### 4.8   Blockchain DL in Education Sector

A DL is stored for each student's records in the educational sector, and if a student loses any information, it can be restored from the DL. The DL link can also be provided with any authority for verification of student information based on student access permissions.

## 5   Security Analysis

The following formal security verification tools and formal security analysis methods are needed for security analysis protocols [28, 29].

### 5.1   Avispa

AVISPA tool is a formal security verification tool. In this tool, the security protocol is written in the HLPSL format. The AVISPA tool validates the security protocol to be secure from the man-in-the-middle (MITM) attack and replay attack. The architecture of the AVISPA tool consists of four backbends such as TA4SP, CLAtSe, SATMC, and OFMC. The security protocol [20] is written through HLPSL language and it is represented by each participant role. After each role of participant, session and environment roles are also represented. AVISPA tool supports cryptographic primitives such as encryption, signature, hash function, and XOR operation. In the environment role, intruder (i) also participates as a legitimate user based on the Dolev–Yao model. The HLPSL is converted to intermediate format using the HLPSL2IF translator. The Intermediate Format (IF) produces output through backbends (anyone). After backbends test, it shows the results. From the results, we know that security protocol is against replay attack and MITM attack or not. The detailed explanations of the AVISPA tool are given in Guide [21].

## 5.2  Scyther

Scyther tool is used for formal security verification of security protocols. In this tool, it is assumed that all the cryptographic primitives are perfect. It is based on like C/java syntax. In the Scyther tool, the security protocol is specified in the security protocol description language (SPDL). In this tool, the security protocol roles of the participant are represented. Each role of the participant has a set of events such as receive, send, declaration, and claim. Scyther tool supports cryptography primitives such as symmetric encryption/decryption, public key infrastructure, hash function, etc. The output shows the claims of the events in the security protocol. The detailed document of the Scythe tool is presented in Cremers [22].

## 5.3  Proverif

Proverif tool is used for formal security verification of security protocol. It supports cryptography primitives such as symmetric encryption/decryption, asymmetric encryption/decryption, digital signatures, hash functions, non-interactive zero-knowledge proofs, and bit commitment. In Proverif tool, the security protocol is written through pi-calculus. It is having three parts such as declaration, process macros, and main process. Proverif tool is to verify reachability, authenticity, and secrecy. Proverif tool is having the Dolev–Yao threat model. In the output of the Proverif tool, it shows three kinds of the results such as queries are true, false, and cannot be proved. A detailed explanation of the Proverif tool is given by Blanchet et al. [23].

## 5.4  Tamarin Prover

Tamarin prover [24, 25] is a formal security verification tool for verifying security protocols [26]. The input of the Tamarin prover tool is the security protocol model. The security protocol model is constructed based on the different roles of the entities involved in the security protocol and with the involvement of the adversary. Tamarin prover supports additional cryptographic techniques such as bilinear pairings and Diffie–Hellman exponentiation. Tamarin prover tool gives constructing proofs in two methods such as fully automated mode and interactive mode. A detailed explanation is given in Li et al. [18].

## 5.5  BAN Logic

BAN logic was proposed by Burrows et al. [27]. The BAN logic is used to analyze the formal security proof of security protocols [22, 23]. In BAN logic, it consists of hypotheses, rules, and proofs. The hypotheses (assumptions) are made to analyze the security protocol and this assumption is constructed based on freshness. The assumptions are used to prove the security protocol to avoid replay attacks. In BAN logic, rules such as message meaning, nonce verification, jurisdiction, freshness, and belief to achieve the goals of the security protocols based on proofs. A detailed explanation of the BAN logic is given by Burrows et al. [27].

## 5.6  RoM Model

Random oracle Model (RoM) was proposed by Bellare and Rogaway [27]. The RoM model [30] is used for formal security analysis of security protocols. It is a black box. The black box gives random response output from its output domain based on the individual query. If the query is repeated many times, the same method responds to the requested query. A detailed explanation of the RoM model is given in Bellare and Rogaway [27].

# 6  Comparative Analysis

## 6.1  Comparative Study of DL-Based Security Protocols

Jangirala et al. [11] designed blockchain-based RFID-enabled authentication protocol for supply chains in mobile edge computing using 5G technology. In their protocol, they designed protocol based on hash function, XOR operation, and bitwise rotation operations. The session key was generated between the supply chain node and Tag (T) with the help of Reader (T). They performed security analysis using informal and formal security verification using the AVISPA tool. The Jangirala et al. [11] protocol does not verify using formal analysis, and the communication cost is high.

Guo et al. [12] suggested a blockchain-based authentication protocol for edge computing. Their protocol is designed based on Elliptic Curve Cryptography (ECC), bilinear paring, and consortium blockchain is adopted for authentication. In their protocol, the consensus algorithm is executed for storing authentication logs and verifying the identity based on the blockchain network. They performed simulations using MATLAB and hyper-ledger fabric. The security analysis of their protocol is performed with fewer parameters and also their protocol does not verify using "formal security analysis" and "formal security verification".

Lin et al. [13] designed a "mutual authentication" protocol for smart homes using blockchain. Their protocol was designed based on public key encryption and group signatures. They used permission blockchain for their protocol based on the PBFT consensus mechanism. The "mutual authentication" was performed between the home gateway and users. The drawback of their protocol: does not have formal security verification.

Odelu [14] designed biometric-based user authentication protocol using consortium blockchain for identity management. Their protocol was developed based on bilinear pairing and ECC. In their protocol at the time of initialization phase Registration Center (RC) and Authentication Server (AS) generates public/private key pairs and joins to the blockchain network with their public keys. In user registration phase, RC generates some secrets for user and RC puts information about user in blockchain network. In authentication phase, the user gives login request with their identity to authentication server and the authentication server will verify the identity with the help of blockchain. After verification of identity, the Session Key (SK) is generated between user and the authentication server for secure communication. In their protocol they used consortium blockchain for identity verification of user. They performed security analysis through informal security analysis. The protocol drawbacks of this protocol are they did not do the "formal security analysis", "formal security verification" and also their protocol takes high communication and computation cost.

Kumar et al. [15] suggested iris authentication using blockchain-based on additive Elgamal homomorphic encryption. In their protocol, they used homomorphic encryption, hash function, and blockchain. In their protocol, they performed authentication using blockchain. In their protocol at the time of the enrollment phase, the encrypted user iris template is stored in the blockchain with user identity information. At the time of authentication, the client device sends the user identity and encrypted user iris template to the blockchain and the blockchain gets the encrypted user iris template from the server. The blockchain computes a hash of the encrypted user iris template and finally, blockchain compares user iris template values and retrieved server values. If both values are the same, then the blockchain computes the distances and it can be sent to the client device for authentication test. The drawbacks of their scheme are it does not provide user anonymity and un-traceability property at the time of the authentication phase, does not provide formal security analysis and does not have formal security verification.

Shen [16] designed Support Vector Machine (SVM) training for blockchain-based Internet of Things (IoT) data in a smart city environment. In their method, they used a homomorphic cryptosystem to secure IoT data and blockchain for storing encrypted IoT data. SVM for data classification and its application are disease diagnosis and anomaly detection. The training and testing phase is performed based on encrypted data stored in the blockchain.

Zhang et al. [18] designed integrity checking data stored in the cloud through blockchain against procrastinating auditors. In their protocol, they used hash function, bilinear pairing, and blockchain. They have five phases such as setup, store, audit, log generation, and check log. In the setup phase, initial parameters are determined by Key Generation Center (KGC). In the store phase, the user (U) outsources their data into the cloud along with some security parameter (SP) (i.e., Data ($D_i$), i.e., $I = 1$ to n, SP). In the audit phase, a third-party auditor checks the integrity of the outsourced user data in the cloud with the help of blockchain. In the log generation phase, third-party auditor generates the log file, and the log file is uploaded into the blockchain. In the check log phase, the user audits the log file stored in the blockchain. The drawbacks of their scheme are they performed security analysis in some of the attacks only and they did not perform detailed security analysis, does not have "formal security verification" and the outsourced data blocks are plain text format.

## 6.2 Quantitative Analysis

Table 1 shows the quantitative analysis for DL-based protocols.

## 7 Future Challenges

- In the blockchain, a distributed ledger is used to solve the centralized system problems, but each node in the network takes huge storage space for blockchain.
- In a blockchain network, the possibility of an attack is 51%. To solve these issues, cryptography techniques are needed for secure data storage. We listed some of the cryptography techniques and also we listed some of the security analysis methods in Sects. 2 and 5, respectively.
- The challenges were discussed in Sect. 6.4. In the literature survey also, we discussed blockchain-based security protocol and also we listed the security challenges of the existing works.
- PKI schemes are possible with quantum attacks, therefore lattice-based cryptography techniques are needed to solve security and privacy challenges.

## 8 Conclusion

In this chapter, we have analyzed various security and privacy challenges of blockchain distributed ledger based applications and we have presented some of the recent security protocols based on blockchain distributed ledger. Solutions for solving security and privacy challenges of blockchain distributed ledger based applications are addressed. The security and privacy challenges of "blockchain distributed ledger" based applications will play important role in upcoming years.

**Table 1** Quantitative analysis

| Author and Year | Cryptography techniques and technology used | Security analysis | Disadvantage |
|---|---|---|---|
| Jangirala et al. [11] | Hash, XOR, bitwise rotation operations and blockchain | Informal, formal security verification using AVISPA | Does not verify using formal analysis, communication cost is high |
| Guo et al. [12] | ECC, bilinear paring and consortium blockchain | Informal | Security analysis is performed with less parameter, Does not verify using formal security analysis and formal security verification |
| Lin et al. [13] | Public-key encryption, group signatures, permission blockchain | Random oracle model, informal security analysis | Does not have verify using formal security verification |
| Odelu [14] | Bilinear pairing, ECC, consortium blockchain | Informal | Does not verify through informal security analysis and formal security verification, high communication and computation costs |
| Kumar et al. [14] | Elgamal homomorphic encryption, blockchain, hash function | Informal | Does not provide user anonymity and untraceability at the time of authentication, does not provide formal security analysis, and does not have formal security verification |
| Shen [16] | Homomorphic cryptosystem, blockchain, SVM | Security proof | Does not have formal security verification and formal security analysis |
| Zhang et al. [18] | Hash, bilinear pairing and blockchain | Informal | Did not perform detailed security analysis, does not have formal security verification, and outsourced data blocks are plain text format |

**Acknowledgements** The authors wish to thank the SRM University-AP, Andhra Pradesh, National Institute of Technology, Tiruchirappalli, and IDRBT Hyderabad and also anonymous Reviewers, Editor, and Associate Editor(s) for their valuable suggestions.

# References

1. Nakamoto S, Bitcoin A (2008) A peer-to-peer electronic cash system. Bitcoin
2. Lao L, Li Z, Hou S, Xiao B, Guo S, Yang Y (2020) A survey of IoT applications in blockchain systems: architecture, consensus, and traffic modeling. ACM Comput Surv (CSUR) 53(1):1–32
3. Zhu Q, Loke SW, Trujillo-Rasua R, Jiang F, Xiang Y (2019) Applications of distributed ledger technologies to the Internet of Things: a survey. ACM Comput Surv (CSUR) 52(6):1–34
4. Stallings W (2006) Cryptography and network security, 4/E. Pearson Education India
5. Kobliz N (1987) Elliptic curve cryptography. Math Comput 48:203–209
6. Menezes AJ (2012) Elliptic curve public key cryptosystems, vol 234. Springer Science & Business Media
7. Maurya AK, Sastry VN (2017) Fuzzy extractor and elliptic curve based efficient user authentication protocol for wireless sensor networks and Internet of Things. Information 8(4):136
8. Dodis Y, Reyzin L, Smith A (2004, May) Fuzzy extractors: how to generate strong keys from biometrics and other noisy data. In: International conference on the theory and applications of cryptographic techniques. Springer, Berlin, Heidelberg, pp 523–540
9. Irshad A, Chaudhry SA, Alomari OA, Yahya K, Kumar N (2020) A novel pairing-free lightweight authentication protocol for mobile cloud computing framework. IEEE Syst J
10. Lumini A, Nanni L (2007) An improved biohashing for human authentication. Pattern Recogn 40(3):1057–1065
11. Jangirala S, Das AK, Vasilakos AV (2019) Designing secure lightweight blockchain-enabled RFID-based authentication protocol for supply chains in 5G mobile edge computing environment. IEEE Trans Ind Inform
12. Guo S, Hu X, Guo S, Qiu X, Qi F (2019) Blockchain meets edge computing: a distributed and trusted authentication system. IEEE Trans Industr Inf 16(3):1972–1983
13. Lin C, He D, Kumar N, Huang X, Vijayakumar P, Choo KKR (2019) Homechain: a blockchain-based secure mutual authentication system for smart homes. IEEE Internet Things J 7(2):818–829
14. Odelu V (2019, June) IMBUA: identity management on blockchain for biometrics-based user authentication. In: International congress on blockchain and applications. Springer, Cham, pp 1–10
15. Kumar MM, Prasad MV, Raju USN (2020) BMIAE: blockchain-based multi-instance Iris authentication using additive ElGamal homomorphic encryption. IET Biometrics
16. Shen M, Tang X, Zhu L, Du X, Guizani M (2019) Privacy-preserving support vector machine training over blockchain-based encrypted IoT data in smart cities. IEEE Internet of Things J 6(5):7702–7712
17. Maesa DDF, Mori P, Ricci L (2019) A blockchain based approach for the definition of auditable Access Control systems. Comput Secur 84:93–119
18. Zhang Y, Xu C, Lin X, Shen XS (2019) Blockchain-based public integrity verification for cloud storage against procrastinating auditors. IEEE Trans Cloud Comput
19. Li L, Liu J, Cheng L, Qiu S, Wang W, Zhang X, Zhang Z (2018) Creditcoin: a privacy-preserving blockchain based incentive announcement network for communications of smart vehicles. IEEE Trans Intell Transp Syst 19(7):2204–2220
20. Kumari S, Karuppiah M, Das AK, Li X, Wu F, Kumar N (2018) A secure authentication scheme based on el-liptic curve cryptography for IoT and cloud servers. J Supercomput 74(12):6428–6453
21. Guide ABS (2006) HLPSL tutorial
22. Cremers CJ (2008, July) The Scyther Tool: verification, falsification, and analysis of security protocols. In: International conference on computer aided verification. Springer, Berlin, Heidelberg, pp 414–418
23. Blanchet B, Smyth B, Cheval V, Sylvestre M (2018) ProVerif 2.00: automatic cryptographic protocol verifier, user manual and tutorial. Version from, 05–16

24. Meier S, Schmidt B, Cremers C, Basin D (2013, July) The TAMARIN prover for the symbolic analysis of security protocols. In: International conference on computer aided verification. Springer, Berlin, Heidelberg, pp 696–701
25. https://tamarin-prover.github.io/manual/tex/tamarin-manual.pdf
26. Bojjagani S, Sastry VN (2019) A secure end-to-end proximity NFC-based mobile payment protocol. Comput Stand Interfaces 66:103348
27. Burrows M, Abadi M, Needham RM (1989) A logic of authentication. Proc Roy Soc London. A. Math Phys Sci 426(1871):233–271
28. Vivekanandan M, Sastry VN, Reddy US (2019, January) Biometric based user authentication protocol for mobile cloud environment. In: 2019 IEEE 5th international conference on identity, security, and behavior analysis (ISBA). IEEE, pp 1–6
29. Vivekanandan M, Sastry VN, Reddy US (2019) Efficient user authentication protocol for distributed multimedia mobile cloud environment. J Ambient Intell Humanized Comput 1–24
30. Bellare M, Rogaway P (1993, December) Random oracles are practical: a paradigm for designing efficient protocols. In: Proceedings of the 1st ACM conference on computer and communications security, pp 62–73
31. Vivekanandan M, Sastry VN (2020) BIDAPSCA5G: blockchain based Internet of Things (IoT) device to device authentication protocol for smart city applications using 5G technology. Peer-to-Peer Netw Appl 1–17
32. Zhang L, Xie Y, Zheng Y, Xue W, Zheng X, Xu X (2020) The challenges and countermeasures of blockchain in finance and economics. Syst Res Behav Sci 37(4):691–698
33. Guo Y, Liang C (2016) Blockchain application and out-look in the banking industry. Financ Innov 2(1):24
34. Gao H, Ma Z, Luo S, Wang Z (2019) BFR-MPC: a blockchain-based fair and robust multi-party computation scheme. IEEE Access 7:110439–110450
35. Vivekanandan M, Sastry VN, Srinivasulu Reddy U (2020) BIDAPSCA5G: blockchain based Internet of Things (IoT) device to device authentication protocol for smart city applications using 5G technology. Peer-to-Peer Network Appl 1–17
36. Abeyratne SA, Monfared RP (2016) Blockchain ready manufacturing supply chain using distributed ledger. Int J Res Eng Technol 5(9):1–10
37. https://www.who.int/docs/default-source/coronaviruse/situation-reports/20200831-weekly-epi-update-3.pdf?sfvrsn=d7032a2a_4
38. Kalla A, Hewa T, Mishra RA, Ylianttila M, Liyanage M (2020) The role of blockchain to fight against COVID-19. IEEE Eng Manag Rev
39. Zhang L (2008) Cryptanalysis of the public key encryption based on multiple chaotic systems. Chaos, Solitons Fractals 37(3):669–674
40. Xiao D, Liao X, Deng S (2007) A novel key agreement protocol based on chaotic maps. Inf Sci 177(4):1136–1142
41. Zhang R, Xue R, Liu L (2019) Security and privacy on blockchain. ACM Comput Surv (CSUR) 52(3):1–34
42. Pearson S, May D, Leontidis G, Swainson M, Brewer S, Bidaut L, Zisman A (2019) Are distributed ledger technologies the panacea for food traceability? Glob Food Sec 20:145–149
43. De Alwis C, Kalla A, Pham QV, Kumar P, Dev K, Hwang WJ, Liyanage M (2021) Survey on 6G frontiers: trends, applications, requirements, technologies and future research. IEEE Open J Commun Soc 2:836–886
44. Velliangiri S, Manoharn R, Ramachandran S, Rajasekar VR (2021) Blockchain based privacy preserving framework for emerging 6G wireless communications. IEEE Trans Ind Inform

# Computation Time Estimation of Switches and Controllers Process on 6G-Based SDN-Cyber Security Forensics Architecture in the Blockchain-Based IoT Environment

**Deepashika J. Rathnayake and Malka Halgamuge**

**Abstract** Improving the security of blockchain-based Software-Defined Networking (SDN) is discussed in recent studies. The blockchain is utilized on the SDN-based Internet of Things to meet some security provisioning challenges, data integrity, and evidence alternation in digital forensics in cybersecurity. Finding a computation time estimation solution is prominent to overcome the computational time complexity of the 6G-based SDN-forensic architecture in a blockchain-based IoT environment. Functionalities of Software-Defined Networking dynamically control the network flow on the 6G-based SDN-forensic architecture. We develop a model to estimate the computational time on the SDN-cyber forensics architecture that works in a blockchain-based IoT environment in the 6G network. Our results demonstrate the total time consumption on the SDN-forensic network is high regardless of the number of IoT devices. We show the time consumption of controllers is higher than the time consumption on switches. Additionally, we observe the impact of the number of IoT devices on time consumption in switches and controllers. Basic sense, forensic-based processing delay affects the total network progress in the SDN-forensic network. The lower scalability of the 6G-based forensic network is not able to perform securely, as the traffic provided is less. As a result, it can have a significant impact on the final productivity. The low scalable forensic network continuously engages in a wasteful activity and thus is unable to fully benefit as a result of latency and scalability concerns. Therefore, it has the potential to significantly affect final throughput. In conclusion, this chapter presents a comprehensive analysis of computation time for future research to find ways to overcome the time complexity of blockchain-based Software-Defined Networking in the 6G environment.

D. J. Rathnayake (✉)
Charles Sturt University, Melbourne, VIC, Australia
e-mail: deepashika.r@gmail.com

M. Halgamuge
Department of Electrical and Electronic Engineering, The University of Melbourne, Parkville, VIC, Australia
e-mail: malka.nisha@unimelb.edu.au

135

**Keywords** Blockchain · 6G · Cybersecurity · Latency · Software-Defined Networking · SDN-forensic architecture · Switches · Controllers · Computational time · Throughput

# 1   Introduction

The network management system in 6G is evolving toward integration, distribution, diversity, and intelligence as information technology advances. The 6G will enable revolutionary applications by merging sensing, imaging, and precise timing with mobility and truly leveraging artificial intelligence and intelligent networks such as forensic software-defined networks. The 6G-based forensic network system will improve 5G in terms of performance and user Quality of Service (QoS) while also adding some fascinating new features in the SDN-forensic network. It will protect the forensic system and user information. It will offer convenient services. The SDN is also seen as a key technology for the 6G network. One of the most significant criteria for a 6G network architecture is flexibility, as well as the improvement of 6G network performance, which SDN technology can achieve. There is a novel network management approach which has a massive network with heterogeneous devices. By decoupling the control plane from the data plan in the 6G network, SDN can overcome the vertical integration.

Blockchain is the solution for SDN that can audit forensically unchanged logs. Blockchain is the leading software platform globally that helps for security enhancement and the quality control perspective of digital forensics. Private key cryptography, shared ledger distributed network, and network service transaction incentives are the three technologies that keep blockchain records securely [1]. In previous studies, secure automation is discussed. However, it is becoming increasingly difficult to ignore controllers' computation time and switches in 6G SDN-cyber forensics architecture in cybersecurity in the blockchain-based IoT environment. The computation time of switches and controllers plays a vital role in SDN-forensics architecture in cybersecurity. Usually, a large number of IoT devices connect with blockchain networks through a gateway.

Estimating computation time allows identifying the time consumption of switches in the data plane and controllers in the control plane. Furthermore, we study the effect of security in signature verification and the digital signature providing secure control ownership. The purpose of this chapter is to review the recent estimation of computation time during the SDN-forensics lifecycle in cybersecurity of the blockchain-based IoT environment. A novel blockchain-based distributed cloud architecture with software defined networking enables controller fog nodes at the edge of the network to meet the required design principles [1, 2]. Cyber Forensics users are validated by using the hash algorithm in the control plane, and data is secured by homomorphic encryption with high privacy in a decentralized model. It protects non-authorized modifications, authenticity, and non-repudiation. The novel blockchain-based architecture provides low-cost, secure, and on-demand access and addresses high availability, real-time data delivery, scalability, security, resilience, and low latency.

An electric vehicle power trading model based on blockchain and smart contract dramatically improves energy efficiency and cost-effectiveness, achieving efficient operation [4]. The SDN-based energy internet is a distributed architecture supported by blockchain that achieves a good match of the transaction object protection [5]. So far, there has been little discussion about the security and privacy of blockchain and SDN. We have found a computational time estimation model on the 6G-based SDN-forensics architecture in a cybersecurity blockchain-based IoT environment.

## 1.1 Motivation

The time estimation of SDN-forensic architecture in a blockchain-based IoT environment is the prominent solution to overcome the computation complexity of blockchain-based 6G SDN. Most of the previous work focused on security issues, network complexity, and latency issues of Software-Defined Networking. To address these security issues, blockchain technology is used. However, there is no systematic model to estimate the computational time on SDN-forensics architecture in a cybersecurity blockchain-based 6G IoT environment. Therefore, our motivation is to propose a model to estimate total time consumption to overcome the high-level complexity of computational time.

## 1.2 Main Contribution of the Chapter

The computational time estimation model proposed in this study analyzes the time consumption of switches and controllers on 6G-based SDN-forensic architecture in a cybersecurity blockchain-based 6G IoT environment. Therefore:

- We propose a model to estimate the total time consumption on SDN-forensic architecture in cybersecurity for switches and controllers on SDN-cyber forensic architecture in a blockchain-based IoT environment.
- We observe the time consumption for switches and controllers with the increment of the number of IoT devices and the amount of data.

The rest of this chapter is organized as follows: Sect. 2 begins by laying out the materials and methodology of the research, and Sect. 3 illustrates the significant achievements of the proposed work, along with the results obtained. Finally, Sect. 4 concludes this chapter by outlining future directions.

## 2  Materials and Methods

The digital signature is the critical control in forensic SDN. The forensic network uses a mathematical algorithm to provide authenticity of the data and protection against forgery. Blockchain allows multiple signatures. Blockchain adds the most value in the concept of consensus; transactions cannot be edited or deleted, which significantly secures transactions and signature technologies. Hence, the security issues related to the blockchain are essential in conditions for cybersecurity. In this sense, security experts need to fully understand the scope and impact of security and privacy challenges related to the blockchain before predicting potential harm from an attack and checking if current technology can withstand constant hacking attempts. There is a threat from hackers that they use many different ways to infiltrate a cyber-forensic network, this could be a phishing attack in which hackers mimic a Wi-Fi network or program and urge employees of an organization to log in and give them their credentials. The hash key is utilized for integrity checking as it is the digital signature. When considering SDN controller forensics architecture in cybersecurity, it develops an algorithm to verify the user. Each user signature validates before getting permission to the data logs, and then it becomes authentic and legitimate. Blockchain is utilized in the control plane, and y has made significant contributions to cybersecurity due to its immutability, traceability, decentralization, and transparency. Encryption and verification are two of the essential parts of a cybersecurity network, and blockchain offers both. In cyber-forensic architecture, the algorithm is running in a blockchain network. Block 1, block 2, block 3, and block 4 are the secondary evidence databases, and each block contains *block_header, block_body, hash, previous_block_hash, time_stamp, forensic_data_set, consensus_algorithm, nonce, version,* and *evidence source.* Once the evidence reaches from the IoT device, its destination is signed. The signature is created using a private key and securely stored by the signer. No matter about any kind of cyber-attack, we use a decentralized database in the forensic network as the blockchain is decentralized by nature, which means there is no single point of penetration for hackers to invade. Also, it mitigates the risk that comes from any single node being compromised. The blockchain is

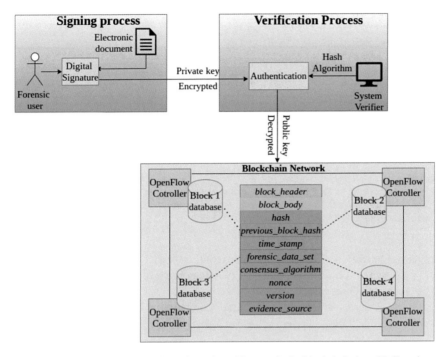

**Fig. 1** Digital signature works in SDN-forensic architecture in the blockchain-based IoT environment with secure access for the evidence database. The forensic user enters the digital signature on an electronic document that is encrypted. After the verification process, the user obtains a block of evidence for the blockchain network

also operating on a multi-signature authentication model, which avoids usernames and passwords in favor of user authentication by confirming that they have access to multiple devices. This is a more secure way to grant network access, ultimately reducing the ease and frequency of breaches and better protecting forensic data (Fig. 1).

## 3  Computation Time Estimation on SDN-Forensic Architecture in Blockchain-IoT Environment

Blockchain technology is utilized in the cyber forensics environment to facilitate modern digital cyber forensics. The Internet of Things is incorporated with digital cyber forensics introducing complexity to cyberspace. Records generated by IoT devices can assist with event reconstruction. The ubiquitous deployment of IoT devices enhances connectivity and communication through the 6G network. Its integrity and thus, its acceptability can only be achieved if the chain of custody (CoC) is maintained as its blockchain within the broader context of the ongoing

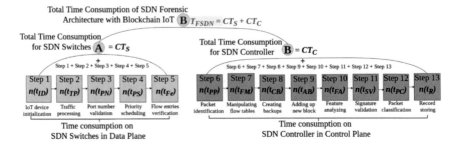

**Fig. 2** Total computation time estimation for SDN-forensic architecture in the blockchain-IoT environment. Step 1 to Step 5 is used to obtain the total time consumption on switches in data plane as Section A. Step 6 to Step 13 is run in the data plane as Section B. Finally, combining Section A and Section B, the total computational time of SDN-forensic architecture in the blockchain-IoT environment can be estimated (Fig. 3 corresponds to Fig. 2)

digital investigation. Therefore, the blockchain acts as a controller of lacking data and server compromise of service quality in the digital forensics' environment with cybersecurity 6G network.

Figure 2 demonstrates the total Time Computation Estimation for 6G SDN-forensic architecture in the blockchain-IoT environment:

- **Step 1**: there are n number of IoT devices that come from the device plane and start the IoT device initialization at the data plane.
- **Step 2** identifying network traffic of each data packet allows traffic types of VoIP, HTTP, and FTP.
- **Step 3** the endpoint of logical connection in the 6G SDN-forensic network that is used for specialized services.
- **Step 4**: priority scheduling engages with traffic processing; generally, the packet priority is used only when the service requires consistent high performance, which is extremely important.
- **Step 5**: flow entries verification typically updates the switch with new flow entries while receiving new packet patterns, whereas the switch can handle it locally.
- **Step 6**: time consumption for packet identification, in the blockchain network, which is done by the packet parser, dynamically monitors the packets which are coming from the devices.
- **Step 7**: OpenFlow controllers in manipulating flow tables allow for the control of switching rules because OpenFlow is a control protocol that communicates policies and traffic management information between switches and controllers.
- **Step 8**: the creation of binary backups is essential to store binary images of forensic evidence.
- **Step 9**: a new block is added into the forensic network with a hash key.
- **Step 10**: the feature analysis is used to extract packet features that arrived from the data plane. All packet features are not extracted at this stage, except for specific features.

- **Step 11**: signature validation phase-in, authenticates the blockchain which is running to complete the validation process.
- **Step 12**: packet classification is always done after the feature extraction.
- **Step 13**: controllers store evidence records to data logs with the hash keys for secure access after the packet classification. Then time consumption on switches ($T_S$), controllers ($T_C$), and the total time consumption on SDN-forensic architecture in the blockchain-based IoT environment ($T_{FSDN}$) are represented in Fig. 2.

Figure 3 corresponds to Fig. 2 and is explained as follows. Equation (1) shows the data plane with appropriate steps from Step 1 to Step 5 and Eq. (2) shows the control plane with respective steps from Step 6 to Step 13 on SDN-forensic architecture (Fig. 3). There are four separate layers in the SDN-forensic network, including IoT device plane, data plane, control plane, and application plane. An IoT gateway maintains the communication gap between the IoT device plane and the data plane. We assume IoT devices offer local processing and storage solutions when systematically connected with the data plane. Each block contains an OpenFlow controller to support 6G SDN. The northbound interface is the connection between the controller and the forensic network's applications, and the southbound interface is the connection between the controllers. Blockchain is set for transaction verification where it has become an emerging technology with decentralized, transparent, and immutable features. Clusters of computers manage and distribute a time-stamped series of records of evidence in the 6G SDN-forensic network, which is used by the blockchain network. Forensic users carry out every write operation, and these operations are validated through consensus checks, where the entire network should agree to make the changes. The software-defined network is protected from attacks that provide automatic and programmable rules. The flexibility and resilience of the forensic network can be improved by a distributed control plane. The OpenFlow controllers take routing decisions allowed by OpenFlow switches, push forwarding rules, and security rules on switches to control the network switches. This would help manage traffic in the forensic network. The control plane automatically stores evidence in a forensic database that can be modified according to forensic purposes and can create binary backups. However, the primary evidence database is the main database that stores binary images of evidence. We present Algorithm 1 for the time estimation process and Eq. (3) for time calculation.

Time estimation is based on the IoT environment due to IoT generating vast amounts of forensic data and delivers large amounts of data in real-time. Developing a computation model on SDN is that separating the control plane from the data plane and the network is centrally managed and programmable. Although the Network Function Virtualization (NFV) is a function of 6G SDN, there are some limitations not to use in the forensic deployment computation model. Because NFV environments are more dynamic and might require scaling up with additional features to cope. We are also not supposed to develop the computation model on any SDN-based architecture, and security issues come with routers and switches. Also, the

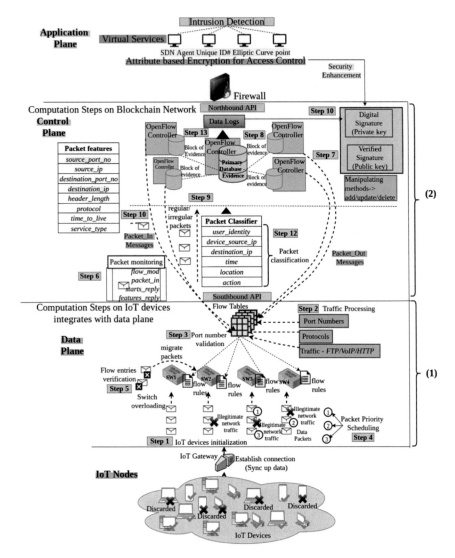

**Fig. 3** Switches and controllers' behavior in SDN-forensic architecture in blockchain-IoT environment. IoT devices send packets to the data plane, then start the operation with the IoT device initialization. After traffic processing is completed, validate the port numbers. Packets reach the switches after the priority scheduling, start flow entries verification process, then packets migrate to the next available switch when the switch is overloaded. Step 1 to Step 5 are in the data plane. After the process of the data plane, packets reach the control plane. Step 6 to Step 13 is explained as follows. Once the packet identification is made, the controllers manipulate the flow tables. Next, create binary backups to store binary images, feature extraction, which is completed by feature analyzer, and validate the signature of the forensic user for authentication purposes. When packet classification is completed, evidence records are stored in the blockchain with the hash key

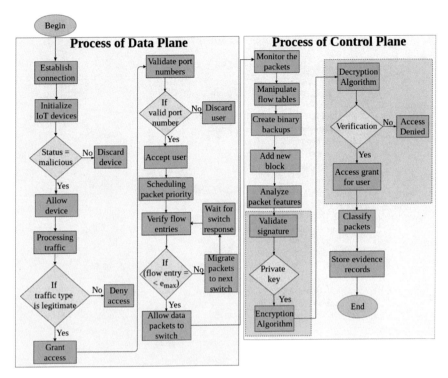

**Fig. 4** Flowchart for switch and controller operations in the SDN-forensic architecture in blockchain-based IoT environment

computation model we proposed is unique, and the steps of the forensic network steps are not similar to any SDN/NFV architecture based on 6G (Fig. 4).

Algorithm 1: Computation time on data plane, and blockchain network plane of the 6G SDN-forensic architecture in blockchain-based IoT environment.

1. **Begin** //*start the process*
2. **Establish** connection //*sync data with data plan*
3. **Initialize** IoT devices *1, 2, 3,.....,n* //*n number of IoT devices*
4.   **If** (Status = True)
5.   **Verify** IoT Device
6.   **If** (malicious traffic)
7. {
8.     Allow IoT Device //*allow access to the network*
9. **Else**
10.    Discard IoT Device //*access denied*
11. }

12.  **End if** *//end for IoT devices initialization phase*
13. **Process** traffic *//identify the legitimate and illegitimate traffics*
14.      **If** (traffic type = *VoIP, FTP, HTTP*)
15. {
16.      Grant Access *//legitimate network traffic types*
17. **Else**
18.      Deny Access *//illegitimate network traffics*
19. }
20.    **End if** *//end for traffic identification phase*
21. **Validate** port numbers *//identify valid port numbers*
22.      **If** (Port number = True)
23.      **Verify** port
24.      **If** (valid port number)
25. {
26.      Accept user *//allow to access- Legitimate users*
27. **Else**
28.      Discard user *//discard illegitimate users*
29. }
30.    **End if** *//end of port validation process*
31. **Schedule** priority *//packets prioritization*
32. **Verify** flow entries *//choose the maximum size of flow entries for the switch*
33.      **If** (Flow entry =<$e_{max}$)
34.      **Identify** flow limit
35.      **If** (maximum limit)
36. {
37.      Accept data packets *//switches allow data packets*
38. **Else**
39.      Migrate *//when exceed the limit, migrate to nearly switch*
40. }
41.    **End if** *//end for switches process on SDN*
42. **Identify** packets *//monitor the packets status – regular/irregular*
43. **Manipulate** flow tables *//use methods for manipulating flow tables on switches- add, update, and delete*
44. **Create** binary backups *//collect binary images of forensic evidence - Primary source*
45. **Add** block – *Adding up new block into blockchain recursively*
46. **Analyze** features *//analyze packet features which arrive from data plane*
47. **Validate** signature *//identify valid and invalid user authentication*
48.      **For** all evidence
49.      If (valid signature)
50. {
51.      **Authenticate** user *//allow access for data logs*
52. Else
53.      **Block** user *//deny user access*
54.
55. **Else**
56.      Block user *//deny user access*
57. }
58.    ***End if*** *//end of the authentication process*
59. ***Classify*** *packets //packet classification rules*
60. ***Store*** *forensic records //save data in the forensic database with - hashes*
61.    ***End for*** *//end for controller process on SDN*
62. ***End***   *//finish the   process   on   Software-Defined   Networking*

Algorithm 1 describes the complete process of SDN-forensic architecture in the blockchain-based IoT environment of the 6G network. If device status is true at the initial stage, it is malicious traffic and allows access to the network otherwise, discards the IoT device. Port numbers belong to *VoIP*, *FTP*, and *HTTP*. At the port number validation phase, the users are accepted who have valid port numbers. Packet prioritization is done using basic scheduling, First Come First Serve-Push Out (FCFS-PO), or First Come First Serve-Push Out-Priority (FCFS-PO-P) scheduling. The maximum size of flow entries is identified at the flow entry verification phase. Both regular and irregular packets are identified using a packet parser, then manipulated flow tables in the next stage. After creating binary backups, a new block is added to the blockchain with a timestamp. Thereafter, the packet features are analyzed. At the signature validation, the valid and invalid user access is identified. The packets are classified according to packet features at the next stage and finally stored in the evidence records after the modification.

## 3.1 Computation Time of Data Plane in Which IoT Devices Work

Attacks targeting southbound API and data plane components are included in the SDN-forensic network. We identify device attacks, protocol attacks, and slide channel attacks. The most common attack in the forensic network is a device attack. A vast number of devices reach the data plane. Therefore, a device attack is a major threat to the 6G-based SDN-forensic network because it refers to all attacks. The device attack targets software or hardware vulnerabilities of switches to compromise the SDN data plan. In the simple term of cybersecurity, an attacker may target forwarding devices' software bugs or hardware features. OpenFlow allows direct access to the infrastructure or redirection level of a cybersecurity forensic network with software to allow manipulation and control of its operation. With OpenFlow, provide control of all elements, including switches and other devices at the infrastructure level. There are some forwarding policies that are associated with the data plane, and they are dependent upon the type of device. These devices can be either collocated or dislocated. At the same time, both types of devices might reach the forensic data plane. The collocated devices can be either physical or virtual, and these devices use distributed control planes. The dislocated devices distribute across multiple elements that are centrally controlled. OpenFlow enables the control plane to define the required forwarding behavior of the data plane in a centralized manner. The specified network control policies are reflected by traffic forwarding decisions and are translated by controllers into actual packet-forwarding rules which are in flow tables of OpenFlow switches.

In more specific terms, a flow table and an OpenFlow secure channel consist of OpenFlow switches for external OpenFlow controllers. Flow entries are maintained by forwarding tables, and each flow entry compromises match fields containing header values. The switch processes all incoming packets compared against flow tables that match the entries with packets based on priority order which the controller specifies. If a matching entry was found, increasing the flow counter and the procedures associated with entering a specific flow entry are performed on an incoming package that belongs to the flow counter. According to the Open flow specification, there are several actions including forwarding a packet out on a specific port, dropping the packet, removing, or updating packet headers. The SDN-forensic network links the advantages including flexibility and innovation to network management. There are some issues such as scalability and performances that mainly all forensic network intelligence and control the logic to OpenFlow controllers, hence restricting the OpenFlow task to dumb execution of redirect actions. DevoFlow, which is an OpenFlow rule, claims to reduce switch interactions to the controller by introducing new control mechanisms inside the switches. In this way, local control decisions can be taken by switches while dealing with recurring events, without engaging controllers whose primary tasks will be limited to maintaining centralized control over a much smaller number of important events that require network-wide visibility. The forensic network uses OpenFlow switches using additional state tables to reduce reliance on remote controllers for applications involving local states such as MAC learning processes and port knocking on firewalls.

The OpenFlow protocol is the core of the 6G-based SDN technology, SDN with OpenFlow key promises flexibility and the forensic network's rapid configuration. It is important to note that the forensic network has a dynamic environment that continually changes its evidence and decision-making process. Therefore, OpenFlow protocol is the programmable protocol that is used by the forensic network as it separates the programming of network devices from the underlying hardware and offers a standardized way of delivering a centralized, programmable network that can quickly adapt to changing network requirements. Therefore, the OpenFlow switch enables data to communicate over the OpenFlow channel to an external controller. The packet lookup and forwarding are performed following one or more flow tables and a group table. The SDN-forensic architecture in the 6G network has functioned with three essential elements that are flow tables installed on switches, a controller, and a proprietary OpenFlow protocol for the controller to talk securely with switches. Controllers impose policies on flows through switches and paths designed by controllers on OpenFlow switches to provide traffic management consistency. Enabling OpenFlow on physical keys and migrating to an OpenFlow key is something that most clients are working toward. We assume that the 100 GB switch supports OpenFlow 1.3, which OpenFlow switches in the 6G-based SDN-forensic networking environment. The packet-forwarding plane is disaggregated from the control plane, and switching decisions are made by the forwarding plane. OpenFlow aims to enable an open-source network architecture. There are some benefits due to the decoupling of the data plane from the control plane, including improvement of traffic management capabilities.

## (1) Time for Initializing IoT Devices ($t_{ID}$).

IoT device initialization is the process of identifying devices on the software-defined network. The Internet of Things (IoT) is becoming an increasingly attractive target for cybercriminals. Furthermore, IoT devices are not high-power devices that have minimal embedded security solutions. Therefore, IoT devices gain access to the forensic network after providing identities appropriately and accurately. These device identifiers are closely related to its features, either hardware or software. However, the IoT plane does not reside in the middle of the SDN-forensic architecture of the 6G network as it is the initial point of the network that the entire forensic network depends on the IoT device plane. IoT devices work individually and independently in the forensic network. Many illegitimate device attempts are neglected by themselves. Alongside, IoT devices fail to handle vast amounts of attacks at the same time due to a lack of computing capabilities to handle these attacks. IoT devices use network scanning for particular attacks because certain IoT ports open all the time. Each IoT device in the forensic network can consist of a sensor, actuator, and communication infrastructure.

As we mentioned before, some more devices are identified by using explicit identities, including IP address, MAC address, and other network identities. In the SDN-forensic network, IoT devices are classified as resource-constrained devices that have the processing power, communication capability, memory, and energy when there are some complex cryptography algorithms used for the security of device identity. We identify two phases in IoT device initialization. IoT devices that are connected to the forensic network are first authenticated. These IoT devices use a lightweight authentication protocol, as the IoT devices are resource-constrained. The sharp rise in the use of Internet of Things (IoT) devices has imposed new challenges in device identification due to a variety of devices, protocols, and control interfaces. Also, all devices are heterogeneous with different operating systems and connectivity capabilities. In the second phase, the forensic network determines that service requests from authenticated IoT devices are authorized.

Connectivity, interoperability, and dynamic composability to facilitate communication, data flow, device management, system customization, and service personalization are offered by the IoT device platform. Also, it integrates with all other layers; data plane, control plane, and application plane to streamline infrastructure management and support security at multiple points across the IoT stack. Then data packets are forwarded to the switches, and the algorithm is executed in a loop until all n number of IoT devices are identified. An increase of malware-loaded IoT devices is a threat for the Software-Defined Network and filters malicious traffic by IoT device initializing.

## (2) Time for Traffic Processing ($t_{TP}$).

In the traffic processing stage, forensic data is moving across the network. The important component for network traffic measurement, control, and simulation is network traffic. Certain flow rules are used by switches that are present on the data plane that are generated based on the type of traffic, protocol, and port numbers.

Traffic is classified with a more advanced monitoring approach following flow's importance. Traffic types can be identified as VoIP, FTP, and HTTP in the forensic architecture and generated on switches in the data plane. The IoT devices are allowed only with these traffic types and unnecessary illegitimate network traffic is ignored. Network traffic processing is one part of security analysis and it is a critical element in the 6G forensic network as it detects network threats earlier. More advanced monitoring approaches try to classify traffic according to the flow's importance. Then it takes time for this process. There is specific maximum traffic for each traffic type and when exceeds the size of relevant traffic, which is suspected.

### (3) Time for Port Number Validation ($t_{PN}$).

Network ports are standardized connected devices, and each port is assigned a number. Most ports are reserved for certain protocols, and messages go to the relevant port. A port number is always associated with an IP address of a host, and the type of transport protocol used for communication and specific port numbers are reserved to identify specific services. The ports are closed when required transactions are completed. This provides another layer of security by not leaving these ports open for attack. The flow rules are majorly defined from the port numbers for each traffic. There are different Internet-based apps such as email, web browser, and cloud storage drivers connected with the forensic network. Each of these applications has one or more port numbers. The forensic network allows VoIP, HTTP, and FTP, respective port numbers are 5060, 80, and 21 and they are unique identifiers over a forensic network by specifying both the host and the service. Generally, a port number is associated with an IP address of a network host and the type of transport protocol used for communication in the forensic network.

Port numbers provide firewall security by stipulating the destination of forensic information on the network. We assume that IoT devices within both the internet and the internet access the forensic network through IoT device users' web servers via 6G, by the way, a firewall can be set for network security purposes and then prohibit other packets destined to port 5060,80, and 21 from passing through switches. In the port number validation, reserved specific port numbers to identify specific services. Then an arriving forensic data packet can be easily forwarded to a running application. We suppose that the algorithm runs for each traffic type, then discards the illegitimate users. In simple terms, if any software requires communication with a forensic network system, it will expose a port for the particular software. Allowed port numbers are already identified in the forensic network system, and they are well-known port numbers. Running an algorithm of port number validation for each traffic takes time to proceed and cannot be ignored.

### (4) Time for Priority Scheduling with Multiple Switches ($t_{PS}$).

We assume that there is at least one flow table in an OpenFlow switch. The first flow table matches the incoming packets. The next step is done following the current stage and forms the basic structure. The next scheduling method is FCFS-PO. If the buffer is full when a packet reaches the buffer, it is put at the tail of the queue while the packet at the head is pushed out. Then all the packets move forward in one position. The number of waiting packets decided the position of a packet in the buffer. The last scheduling method is FCFS-PO-P and the process of both arrival and incoming packets are the same as the previous scheduling method. The newly incoming packet has the highest priority, and it will be put in the front position to get service first. Packet prioritization depends on an algorithm that shows log end-to-end data transmission delay, high energy consumption, and deprivation of high priority real-time data packets. Improper allocation of data packets to the queue is the result of this process. The algorithm is static for every priority scheduling method that does not matter on the forensic network's changing requirements.

In simple terms, the purpose of packet scheduling is to select which packet to be dropped or serviced. Therefore, forensic users ensure that the forensic data packet is delivered based on priority and fairness with minimum latency. Also, it ensures the Quality of Service (QoS) of forensic data which in turn increases the transmission rate. There are several network parameters in the 6G-based SDN-forensic network such as bandwidth, packet arrival rate, packet deadline, and packet size. The data packet servicing and dropping are based on these parameters. Therefore, packet scheduling time is based on these parameters. However, time needs to be considered based on which schedule is selected. Therefore, priority scheduling is a long process, together with these three scheduling types.

*(5) Time for Flow Entries Verification ($t_{Fe}$).*

Flexible routing control is enabled by transferring packets according to the flow inputs on the switch. There are a specific number of flow entries for every switch; meanwhile, the verifier chooses the switch's maximum size of flow entries. When packets arrive at a switch during network operation, the arrival packet flow is compared to the flow entries in the flow table. If no match is found, the switch will contact the controller to update the flow table with entries that allow the packet to reach its destination. In the forensic network, medium access control source and destination address, ethernet type, Internet protocol source–destination address are packet meta information, and matching fields are used to match this information. In the flow entry verification, flow entry installation is the critical role, and time for flow entry verification is considered based on every task relative flow entry verification. Then there are two modes of flow entries installation as reactive and proactive mode. In case, the occurrences of failure or change of flow table rules, the flow table is to be updated. Then old flow entries need to be deleted, and new flow entries need to be installed. Apart from that, the flow entry update operation is complicated and takes a long delay due to the flow table reaching its highest capacity. Initially, the primary flow table stores flow entries with timeout calculated by the controller using flow entry knowledge. The secure cyber-forensic SDN environment is maintained through flow entry verification in a forensic 6G SDN. We assume that the running time of flow entries can vary. The flow table is relatively small when compared with the number of required flow rules. There are some performance and security issues in the flow tables, and majorly attackers may overwhelm the flow table with multiple Denial of Service (DoS) attacks. Also, legitimate flow entries will be refused by manipulating the controller.

Time consumption on switches of SDN is given by

$$T_S = n(t_{ID} + t_{TP} + t_{PN} + t_{PS} + t_{Fe}) \qquad (1)$$

### 3.2 Computation Time of Blockchain Network Layer Which SDN Controllers Work

The SDN-forensic architecture does not specify the SDN controller's internal implementation, and it is the strategic control point in the 6G SDN-forensic network. The most fundamental building entity in the 6G-based SDN-forensic architecture, the control plane contains distributed software controllers which handle the communication through forensic network devices and application via open interfaces. SDN controllers play a significant role in the forensic network as clients of the blockchain, demonstrating excellent network scalability. Once SDN controllers install the flow in switches, associating network states are recorded. SDN controllers manage flow control to switches via southbound APIs and the application plane via northbound APIs. The SDN controller receives instructions and requirements from the SDN application layers as it is the logical entity of the SDN-forensic network and relays them to the networking components. The controller extracts forensic information once received from hardware devices and communication back to the SDN application plane, which contains virtual services including statistics and events. In the specific term, application layer requirements are translated by SDN controllers which underline data plane elements and provide necessary forensic information up to SDN applications. The network control logic is supported, and the application layer abstracted view is provided by the SDN control layer, which is referred to as the Network Operating System (NOS). Apart from that, it contains enough information to specify policies while hiding the implementation details of the forensic network architecture. The forensic network's infrastructure layer is associated with IoT devices with basic network functionality for handling and forwarding data packets based on decisions given by the SDN controller.

Blockchain technology solves the problems of consensus and synchronization of multiple distributed SDN controllers in the SDN-forensic network. Each controller gathers OpenFlow commands, and forensic messages are digitally signed with an authentication code to ensure integrity and authentication. Later, it issues a consensus to the third-party blockchain system. The access of validated and unvalidated blocks is verified by controllers and forwarded to blocks to the remaining controllers. There is a heavy workload in the forensic network, as many actions will be carried out at the same time. Therefore, forensic SDN controllers help to simplify network management and reduce the workload of configuration. However, controllers must make sure the information it receives is trustworthy. The SDN controller uses OpenFlow protocols to communicate with switches. The main purpose of the controller is for path calculation. Further individual users are submitting the messages, and controllers are identifying the packets at the initial stage. The message is formed as packet $=$ < *sourceIP, destinationIP, sourcePort, destinationPort, protocol* > in the absence of an existing flow rule. Controllers are responsible for managing the resources with fine-grained access by encrypting each resource with a set of related attributes. The data packets will reach the controller through switches and gateway. For each forensic data stored in the forensic database server, the SDN controller creates a block, distributes it over the blockchain network, and supports evidence collection. SDN implementations allow users to access network resources, deploy new rules, and manipulate network behavior by interacting with a control plane. There may be some abnormal behavior flowing between blockchain nodes and controllers. Then, it immediately blocks the attacking node, such as IoT devices with malicious traffic.

The SDN controller efficiently works and controls both physical and virtual switching in forensic architecture, as well as provides comprehensive network monitoring. However, the control plane is logically centralized and implemented as a physically distributed system. There are two types of controllers that can be used in software-defined networking, such as centralized and distributed. When distributed controllers are compared to centralized controllers, there are some advantages of centralized controllers such as scalability and high performance during the increment of demand requests. As the key component of a software-defined network is the control plane, it ensures smooth management and operation of the entire forensic network. The SDN single centralized controller offers network resilience and availability; however, it is likely to cause single-point failure. Therefore, distributed controllers are recommended for the forensic network rather than ensuring network resilience and availability. We consider that control logic solves the specific networking issues and adjusts the network policy's particular network policy and provides a facility to change the configuration on the SDN-forensic network. Multiple controllers are enabled in the forensic system's blockchain control network to communicate with each other and exchange forensic data over the network, and it is the distributed control configuration. Despite numerous attempts to standardize SDN protocols, there has not yet been any East–West API standard that remains every controller vendor's property. Additional protocols are not required at the data-store level of the SDN-forensic network; however, communication happens there. Therefore, the standardized communication interface provides more comprehensive interoperability between controllers in the SDN-forensic network. Simple high-level policies are allowed by the SDN-forensic network to modify the forensic network as the device level dependency is eliminated to some extent. The controller is designed as a single software console as it facilitates viewing the entire forensic network globally. The forensic SDN controllers introduce new functionality or programs. Nevertheless, it just places them in the centralized controller. There are various controllers with their pros and cons. In the 6G-based SDN-forensic network, we develop a model based on distributed controllers as it brings fruitful results for the forensic requirements. On the other hand, the control plane contains the server as it is the distributed SDN, and it performs the controller's task.

Logical functionalities are implemented on controllers as it is the brain of the SDN-forensic network architecture. Forensic SDN controllers perform various tasks that are really based on forensic investigation. They build flow entries inside the routing devices and keep track of forensic data packets. The flow table is similar to the routing table as flow entries are stored in the SDN controller flow tables. Matching rules, action, and counter are three portions of the flow table. Matching rules include the set of various fields of header portions such as source IP and destination IP. Then, action compromises of operations that perform on the packet processing include packet forwarding to its destination port, packet dropping, and more. Each packet's default action is forward to the controller as the switch does not contain the entry for the relevant flow. Therefore, packets are passed toward the controller. Then controllers begin the packet processing task once it is done and send them back to the switches along with the flow entries. The forensic information which is passed through the network can be grouped based on flow, table, port basis. The controller decided to flow rules in the switch forwarding table. As previously mentioned OpenFlow is the most widely deployed southbound standard from the open-source community. The event-based messages are generated by the OpenFlow controller when it is needed.

### (1) Time for Packet Identification ($t_{PP}$).

The controller's first step is packet identification, which dynamically monitors the packets from devices by packet parser as the forensic data is transmitted along with the network as packets. At the same time, several packets reach the control plane, and controllers will look at the destination address in the header and compare it to their lookup table to find out where to send the packet. There are two phases in jacket identification: identifying fragmented packets and identifying when the sender transmits individual packets. The Control plane combines the identification with the source address to identify a packet uniquely and uses these unique identifiers to reassemble data from packets. However, the IP ID value is specific to each packet, not specific for the entire forensic network. Although data packets get fragmented for packet identification, the IP ID number is the same for all. It is easy to identify switching loops in the forensic network due to the same IP IDs' in the same packet capture.

Fragmentations travel to their destination and where they are reassembled. The packet identification is a 16-bit value used to identify all fragments of a data packet, allowing the destination host to perform packet reassembly. However, In the packet transmission, some packets were dropped, and the IP packet identification numbers facilitated the identification of which packet was dropped. Each packet contains the same attributes: header, payload, and trailer. The packet includes Flow_Mod, Packet_In, Stats_Reply, and other necessary packet features, and we assume that many packets with various sizes are exchanged in the SDN-cyber-forensic network, and the running time varies depending on the packet length.

### (2) Time for Manipulating Flow Tables of the Switch ($t_{MF}$).

It is essential to manipulate the flow table at the SDN switch effectively. Flow information is stored and managed by the flow entry unit in the flow table. OpenFlow switches have flow tables and packet-forwarding information that they obtain from a controller, and the controller in software-defined networking manipulates these flow tables. There are two types of flow tables that are used in switches as hash-based flow tables and wildcard-based flow tables. The hash-based flow table requires a large memory capacity while the wildcard-based flow table utilizes wildcards to store large amounts of flow information in small memory. Although the flow entry search speed is fast in hash-based flow, entry searching speed is slow in wildcard-based flow tables. Therefore, each type of flow table has its own processing time and processing speed to manipulate in its own way. The controller is responsible for maintaining the cybersecurity forensic network state in real time. Typically, a flow table is populated with rules or policies ex: quality of service (QoS), access control lists (ACLs), and IP route tables for fast-forwarding. We can identify the two various occurrences where this operation is done. The first one is reactive as the controller receives a packet from the switch, and the second one is proactive according to the implementation of the OpenFlow controller. Manipulating flow tables is compulsory because it writes legitimate flow entries.

### (3) Time for Creating Binary Backups ($t_{CB}$).

Original data considered as binary backups, therefore, are not used in the analysis phase, whereas secondary data is used. In the software-defined forensic network, binary backups create guesswork and complication in the process. The binary images are used due to the reason that they are the replica of the original data. SDN-forensic network consults the backup controller only when there is a suspicious attack on its primary controller. There is a possibility of missing important data in standard backups, and hence binary backups are used to avoid this issue. Backup data stores ensure that each master controller has a view of the whole network. These data stores can save the forensic network state during the failure time. The backup controllers guarantee the resilience of the control plane with minimum cost. Software-defined backups automate backup and restore while enabling the ability to perform restore independently and on-demand. We consider the software-defined backup environment as it requires individual deployment, configuration, monitoring, and provisioning. We assume that the time is different for each backup due to the size of the original data file. It takes a very long time for and restoring backups for a large database and that could make forensic backups useless.

There is a possibility that the backup path may fail before the primary path due to dynamic network updates. Therefore, when the link fails, the proactively configured route will not be available to route packets on the alternate path. Furthermore, the binary backups may fail earlier than the binary images. As a result of this, the performance is affected, because the incoming data packets are matched with the flow rules, because of the redundancy of backup flow entries. Apart from that, the Backup algorithm is running and uses fewer processing cycles on dedicated controllers, and more parallel backups can be taken, more often. Creating binary backups is not a simple task. Some steps need to be followed to complete the creating binary backups: configure forensic data file, configure sender processes, configure authentication, and create backups. As we see, the whole process of creating binary backups is more complicated.

### (4) Time for Adding Block ($t_{AB}$).

All nodes in the forensic blockchain network are involved in the verification of freshly mined blocks. In simple terms, block time is a measure of the time it takes to produce a new blockchain network block. Similarly, new forensic data needs to be added into the 6G-based SDN-forensic blockchain for the investigation decision process as new forensic information is found in a timely manner. It's basically the amount of time it takes a blockchain miner to find a hash solution, which is a random string of characters associated with a block. However, the exact time to mine the next block is unknown. The actual amount of time it takes to create the block varies depending on the difficulty of hashing. Thus, block time is the average time it takes a miner to solve a math puzzle and build a blockchain block. Therefore, a newly mined block is added to the forensic blockchain based on a consensus algorithm, and it becomes publicly available for anyone to view. The new transaction means a new block that contains transactions and miners validate new transactions. The last block is mined every 10 min, thus adding these transactions to the blockchain. Then transactions become part of a block. Thereby new blocks allow the new owners, and the blocks are considered as confirmed.

Any new transaction added to the blockchain will change the state of the ledger. New transactions are constantly processed by the miners in new blocks added to the end of the chain. However, each block contains its current time. We identify seven steps for adding new blocks into the blockchain as transaction data, chaining the blocks, creating hash signatures, and deciding whether a person can sign the block once the signature is qualified, making the blockchain immutable, governing blockchain, and determining rules. The time is considered based on all the steps. In simple terms, whenever any new block is added to the blockchain, numerous nodes within the same blockchain implementation are required to execute algorithms to evaluate, verify, and process the blockchain block's history. It's common for any blockchain network to include SDN-forensic blockchain. If most of the nodes authenticate the blockchain record and signature, the new block of blockchain transactions will be accepted into the ledger, and the new block containing the data is added to the blockchain. Each blockchain block consists of certain forensic data in the forensic network, the block hash, and the hash from the previous block.

From a cybersecurity perspective, there is an additional level of assurance that the forensic data block is authentic and has not been tampered with. Then, the time for adding blocks increases with the number of IoT devices. The validity of the blocks plays a major role in the cybersecurity forensic network. All the forensic network nodes will check the mined block and add them to the blockchain if it follows the rules stated by the consensus mechanism. The SDN-forensic network simply rejects an invalid block. During the normal operation of the SDN-forensic network, authorized users pay attention to the data logs and periodically check the forensic data blocks' status to ensure the performance of the blockchain network. A blockchain-based log system confirms the forensic network's security as it avoids log tampering by sealing the forensic logs cryptographically. Thus, adding the data logs to a hierarchical ledger, hence, provides an immutable platform for data log storage.

## (5) Time for Feature Analyzing ($t_{FA}$).

Forensic data packets are used in forensic investigations after being captured, stored, and processed efficiently and provide admissible evidence against a suspect in a court case. In the SDN-forensic network, packet analysis is a primary traceback technique. It provides sufficiently detailed packet details and playback even the entire network traffic for a particular point in time. The purposes of feature analysis in the forensic network are finding traces of nefarious online behavior, unauthorized user access, data breaches, malware infection, intrusion attempts, reconstructing forensic data files, etc. Network devices communicate using protocols. These protocols establish connections, and format rules and conventions for data transfer in the forensic network. Packet analyzers do packet feature analysis. These software tools intercept and log network traffic over a digital network or part of a network through packet capture. The utilization of packet feature analysis to its full potential relies on full packet capture,

We identify the message features such as *source_port_number, source_IP, destination_port_number, destination_IP, header_length, protocol, time_to_live,* and *service_type.* Each packet feature is analyzed with proper evidence, and the feature analysis of messages is performed by running the algorithm providing the final output. In practice, time consumption may not be performed by an individual feature, where all features need to get the total duration of feature extraction.

Features are extracted by a feature analyzer and therefore, no difference between feature extraction and feature analyzing. A feature extraction algorithm does feature analysis. We identify two stages of feature extraction as primary feature extraction and secondary feature extraction. In the primary stage, the subset of features is extracted from the dataset having the highest percentage of the target population. Features that can differentiate the target and nested classes are added to the subset of features in the secondary stage. Similarly, feature extraction is an essential component in the forensic network's anomaly detection, summarizing network behavior from a packet stream. Forensic data gathering in high-speed links is a complex process, and therefore existing techniques analyze a small number of packet features and limit their effectiveness. More forensic information is available once the feature extraction is done at the packet level. Besides, a single packet does not offer much information at the packet payload.

## (6) Time for Signature Validation ($t_{SV}$).

The algorithm is running for signature validation and for the computation of network paths and slices. The algorithm runs linearly making all steps to submit at each entry. If user input matches with the system, the signature becomes valid and allows the user. The logger needs to have a keypair as the private key is used to encrypt the forensic data, and the public key is used for decryption. There are several steps in the signature validation phase, such as creating digital signatures, signing forensic messages with private keys, and verifying the forensic message with a public key. These steps took a considerable amount of time to proceed. At this stage, the cryptographic value of the signature is checked using signature verification data. Signature validity is determined by validating the authenticity of the signature's digital ID certificate status and document integrity.

Authenticity verification confirms that the signer's certificate or its original certificates are in the validator's trusted identities list. Thus, it confirms the signing document's validity based on the user's Acrobat and Reader configuration. The document integrity check confirms whether the signed content has changed after it was signed. If the content changes, a document integrity check confirms whether the content has changed in the way the signer allows. There is a signature panel that displays information about each digital signature in the forensic document. Each digital signature contains an icon to identify the verification status. Verification details are listed in each signature and can be viewed by expanding the signature. The signature panel provides information about the signed time and the signer's details. Similarly, a digital signature verifies the integrity of a forensic data file, and it is non-repudiation. The reason is to use the digital signature in the 6G-based SDN-forensic network as it can well adapt to its characteristics in the blockchain system. It will be more secure, applicable, and has the potential to increase value.

### (7) Time for Packet Classification ($t_{PC}$).

Packet classification is the process of categorizing packets into flows. In the SDN-forensic network, all packets go to the same destination equally. Packets are classified based on their headers, and they are classified into flows that are searched in flow tables. However, providing premium services to different users based on their quality requirements has become an issue that requires more requirements. All packets belonging to the same flow obey predefined rules and are similarly processed by the switch. For this, routers should be capable of distinguishing and isolating traffic belonging to different flows. This ability to determine packet flow is called packet classification. The forensic network uses an algorithmic solution for packet classification as it provides a deterministic performance, support for dynamic updates, and added flexibility for the forensic network as it takes advantage of the availability of cheap.

The forensic network's key task is to run an algorithm for packet classification and handle large databases of classification rules. The primary task of routers is to forward packets from the input links to the appropriate output links. As packet classification arose out of the router's need to classify traffic into various streams, the longest-running prefix matching techniques used in routing lookup tables formed the origin of packet classification techniques. The packet classification searches in a rules table for the highest priority rule or set of rules that match the packet. Packet classification is necessary to facilitate packet filtering for security reasons - firewall, packet delivery within specified delay limits (QoS), enabling premium services, policy-based routing, traffic and policing rate limitation, traffic shaping, and billing. Also, packet classification supports high network throughput in the forensic network. Packets are discriminated against, and many differentiated functionalities are enabled by packet classification. There are five tuples from packet headers used for classification: the protocol, destination and source ports, and source and destination addresses. There are two main metrics for packet classification that are speed in-memory access and memory. Packets are classified by source, destination ports, address, and protocol type.

Usually, packet classification is applied in the forwarding plane; however, we assume this process is applied in the controller plane due to its complexity, and blockchain is running on the controller network. At this stage, the context of the packets is identified, and important actions are performed. Some actions identified at this phase are dropping unauthorized packets, scoping, scheduling, prioritizing, and encrypting secure packets. We identify packet classification attributes such as user-identity, *devise_source_IP, destination_IP, time, location,* and *action.* The controller enables the communication of user authentication. However, the device which produces fake identity misleads the network resulting in the high-level packet feature complexity. Changes in packet attributes can cause a significant amount of time consumption.

### *(8) Time for Storing Records ($t_R$).*

Record storing and report generating are an essential part of the forensic software-defined network. All forensic records should be considered the evidence and should be fully accountable in the chain of custody. Likewise, documenting samples is vital, from the first point of entry into the investigation. Once the investigation agent uses the forensic data for his decision-making process, then transfer the updated forensic information and generate a report at this stage. Once forensic users do the modification for investigation purposes, the evidence is sent to either database or generated reports. In this stage, some objectives need to be achieved: ensure confidentiality and integrity of forensic evidence during its modification and storage time, ensure the evidence is collected from a secure system, and compute a non-repudiated proof of the existence of the forensic evidence.

Each record takes a specific time period for its own processing. The logger generates evidence log events and at the time, newly updated evidence materials with forensic results being inserted into the data log. The final output of the 6G-based SDN-forensic network is producing forensic reports, and we consider them as data or evidence logs. The time complexity of report generation is high when volumes of data increase with more complex requests. The total computation time on the controllers during the forensic cycle is given by:

$$T_C = n \left( t_{PP} + t_{MF} + t_{CB} + t_{AB} + t_{FA} + t_{SV} + t_{PC} + t_R \right) \qquad (2)$$

We estimate the computational time of controllers and switches considered to be the best method to adopt in this investigation. Equations (1) and (2) are obtained at processing times, where the total time consumption on the switch is $T_S$, and the total time consumption of controllers is $T_C$. Therefore, the total computation time on SDN-forensics architecture in a cybersecurity blockchain 6G-based IoT environment, $T_{FSDN} = CT_S + CT_C$, is given by:

$$T_{FSDN} = n\,(t_{ID} + t_{TP} + t_{PN} + t_{PS} + t_{Fe})_+$$
$$n\,(t_{PP} + t_{MF} + t_{AB} + t_{CB} + t_{FA} + t_{SV} + t_{PC} + t_R) \tag{3}$$

## 4  Simulation Results

### 4.1  Simulation Setup

A detailed comparative study is presented on the following metrics: computation time
on the data plane, computation time on the control plane associated with blockchain,
and total time consumption in the forensic network. These parameters are significant
for validating the achievements of the SDN-forensics in the cybersecurity blockchain-
based IoT environment.

All parameter values used in our time estimation (for both simulation and the analysis) are listed in Table 1. Since our results are based on IoT device parameters, they
are expected to estimate the real computational time on SDN-forensic architecture in a
cybersecurity blockchain-based IoT environment. In our simulation, we consider the
SDN-forensic network with $n = 500$ number of IoT devices. Consider our developed
model as the square $\{(0,110), (100,115), (200,117), (300,118), (400,119), (450,120)\}$
as in [6]. We generate 500 random setups. Initially, OpenFlow switches are randomly
placed to provide consistency in traffic management and engineering, by making the
control function independent of the hardware it's intended to control. The FS.COM
switch product line consists of 100GbE switch (100G L2/L3 Switch Loaded with
ICOS, 48*25GbE ports + 6*100GbE ports) and supports OpenFlow 1.3, which can
be used as OpenFlow switches in the open networking environment. We generate 500
random setups, each with the following simulation setting. Therefore, each simulation data point is obtained by averaging over 500 random setups. We assume that
the total number of nodes in the entire network $n$ is 500, and each node reports data
once every millisecond.

The results and analysis in this study are performed using MATLAB (MathWorks
Inc., Natick, MA, USA) R20202b on a computer with macOS High Catalina with
Processor 2 GHz Quad-Core Intel Core i5 and RAM (Random Access Memory)
16 GB 3733 MHz LPDDR4X.

We assume that $t_{ID}$ is not a constant value because each device has its own capabilities. There are four major categories of capabilities including transducer capabilities,
data capabilities, interface capabilities, and supporting capabilities. Transducer capabilities include sensing, actuating, and data capabilities include data processing, data
storing, application capabilities include application interface, human user interface,
network interface, supporting capabilities, device management, cybersecurity capabilities, privacy capabilities. However, the initial stage of the SDN-forensic network
is the device initialization. Therefore, total time consumption corresponds to IoT
devices. The time consumption on switches is based on the IoT device initialization

**Table 1** Simulation parameter values

| Symbol | Description | Value |
|--------|-------------|-------|
| $T_S$ | Total time consumption of switches | – |
| $t_{ID}$ | Time for IoT device initialization | 0.02 ms [2] |
| $t_{TP}$ | Time for traffic processing | 4 ms [6] |
| $t_{PN}$ | Time for port number validation | 2 ms |
| $t_{PS}$ | Time for packet priority scheduling | 3 ms [7] |
| $t_{Fe}$ | Time for flow entries verification | 0.5 ms [8] |
| $T_c$ | Total time consumption of controllers | – |
| $t_{PP}$ | Time for packet identification | 0.135 ms [9] |
| $t_{MF}$ | Time for manipulating flow tables | 0.4 ms [10] |
| $t_{CB}$ | Time for creating backups | 1 ms [11] |
| $t_{AB}$ | Time for adding up block | 2.85 ms |
| $t_{SV}$ | Time for signature validation | 100 ms [12] |
| $t_{PC}$ | Time for packet classification | 0.0004 ms [1] |
| $t_R$ | Time for storing records | 0.02 ms |

and traffic processing because IoT devices contain unique identifiers that provide fast access to the network. However, artificial intelligence and machine learning of IoT devices are making the process easier and more dynamic. A firewall is configured not to allow devices with unwanted traffic except $T_V$, $T_f$, and $T_h$, and time for traffic processing increases in the natural order to the number of IoT devices. In the switches, the $t_{TP}$, $t_{PN}$, and $t_{Fe}$ have a mutual connection. Therefore, the time consumption of port number validation and flow entry verification is high in accordance with IoT device initialization.

We consider the time consumption of packet identification is usually high due to the variation of packet size. Every packet needs to be monitored to filter the malicious traffic at the beginning. While processing a packet, switches may check for bit-level errors in the packet that occurred during transmission and determine where the packet's next destination is. When the packet identification time is high, the time consumption of packet processing is high. Thus, we identify the packet processing delay as a constant delay that faces the source and destination. The turnaround time of flow table manipulation is related to the speed at which the controller handles packet in, we assume it takes a considerable amount of time to proceed. In order, $t_{CB}$, $t_{AB}$, $t_{FA}$, $t_{SV}$, $t_{PC}$, and $t_R$ increases in accordance with $t_{PP}$. Further analysis of Fig. 5 shows that regardless of the number of IoT devices, the total time consumption on the SDN-forensic network of 6G environment is high when the time consumption of controllers increases in the natural order to time consumption of switches. Therefore, it is interesting to note that $CT_S < CT_C$ and $T_{FSDN}$ are high (Table 2).

For our simulation, we consider the increase in the number of IoT devices. Note that we should examine controllers because the amount of data transmitted to controllers increases and then, it is a challenge in the IoT environment. The time consumption on switches slowly increases, and therefore, we do not consider

**Fig. 5** The total time consumption corresponding to the number of IoT devices on SDN-forensic architecture. Here we consider n=500 number of IoT devices. This shows the time difference between the switches operation and controllers operation has a significant effect on the SDN-forensic networks total computation time

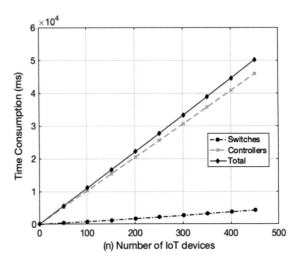

**Table 2** Time consumption in ms

| Number of IoT devices | Switches (ms) | Controllers (ms) | Total (ms) |
|---|---|---|---|
| 1 | 7.7 | 102 | 109 |
| 50 | 395.8 | 5180 | 5576 |
| 100 | 786.4 | 10,264 | 11,051 |
| 150 | 1206.7 | 15,347 | 16,554 |
| 200 | 1654.0 | 20,430 | 22,084 |
| 250 | 2211.9 | 25,518 | 27,729 |
| 300 | 2697.2 | 30,602 | 33,299 |
| 350 | 3162.6 | 35,715 | 38,878 |
| 400 | 3754.3 | 40,805 | 44,559 |
| 450 | 4287.6 | 45,902 | 50,189 |

switches. As in our simulation, the time consumption increment of controllers is high. As a result of this, expanding the network then impacts the latency. Network latency is referred to as block time, and this is the most crucial factor affecting the overall performance of the 6G-based SDN-forensic network. Latency is a big issue in the real world. The reason is latency drives the responsiveness of the network. Therefore, in the forensic network, latency becomes a particular problem.

Network latency, scalability, bandwidth, and throughput are all interrelated components of the SDN-forensic network. However, they all measure different things. There are many IoT devices that need to reach a consensus for a transaction to be verified. The forensic network does not allow any number of data packets at the same time. It has a maximum amount of data that can pass through the network in the given time. It is the way bandwidth integrates with the 6G-based

SDN-forensic network. Thus, there is an average amount of data that passes through over the network, and it's the throughput of the forensic network. According to our results, throughput is not equivalent to bandwidth, because latency affects it. Each node requires access to the entire forensic blockchain. Therefore, it takes a long time for a data packet to be sent to the forensic network's destination. Therefore, it has a high latency network connection. In the data packet transmission, time is the major constraint that causes latency. When data packets get delayed traveling to their destination, the entire network process gets delayed. Latency and scalability are the primary aspects of the SDN-forensic architecture of the 6G network. Network latency impacts the forensic network's scalability, and therefore, controllers are unable to grab many packets. Thus, poor scalability can result in poor network performance.

This latency is a specifically sensitive issue. However, the forensic network will lead to poor quality, if not handled properly. Network latency heavily influences the SDN-forensic network. Also, controllers find it hard to tolerate several IoT devices and amounts of data. Therefore, forensic network processing time is increased. In simple terms, in the SDN-forensic-based processing delay impacts the entire network progress. Less scalability of the forensic network is not capable of performing gracefully, as the offered traffic decreases. The poor scalable forensic network repeatedly engages in a wasteful activity, and therefore, it cannot fully take advantage due to latency and scalability issues. In consequence, it can have a huge impact on final throughput.

Providing access to the entire blockchain network to hundreds of nodes also increases the computation time. We cannot ignore computation time as it has an insignificant high-level time consumption. However, in the 6G SDN-forensic network computation time is quite large, especially where controllers are performing complex encryption algorithms. In the controller's process, data packets are examined for their security which can cause substantial delays in some other steps in the controllers' process. Another point of view, although the switches process has less time consumption than controllers, we cannot decide that the switches process does not have a delay. OpenFlow switches in the forensic network perform network address translation, and there are higher than average processing delays due to switches examining and modifying both incoming and outgoing packets. The forensic network's scalability issue will be the major drawback of the SDN-forensic network as the scalability has a growing demand, which is crucial to the long-term success of the forensic network.

We can apply the observation of this analysis in a real-world scenario. The IoT applications such as vehicular ad-hoc networks and SDN-based 6G network models, where the computation time estimation is essential. These systems require reliable data transmission. Therefore, computation time estimation is a crucial part of these systems, ensuring the time complexity to improve the throughput. Finally, the topic of interest may benefit the vital topic of SDN-based IoT as well as blockchain. Integrating a blockchain with fog computing and IoT to store and compute data at the edge hubs, associated with the off-chain system, may answer scalability issues.

To a certain extent, sharding is a prominent solution [13–16] for making transaction verification in parallel rather than linearly.

## 5   Future Directions

However, more research on this topic needs to be undertaken before the association between switches and controllers is more clearly understood.

### Enhancing for forwarding plane

As the SDN is improved, the control plane becomes more flexible and scalable with new Open-Flow apps. The processing power of software-based controllers can be improved through parallel processing or installing a powerful server's controller.

### Improving manageability

Improving network management using software-defined networks explores the advantage of the software-defined network, simplifying network policy management, in high- evel language, that provides expression and flexibility.

### Languages for Software-Defined Network

High-level abstractions investigate the high-level programming abstractions that need to simplify the creation of SDN applications.

### SDN Platform for Cloud Network Services

The SDN platform explores the SDN application to enable application-level abstractions to communicate in cloud environments.

### Routing and service convergence using SDN

The major challenges for operators in the near future pertain to providing convergent, dynamic, and adaptive networks in the context of a multi-services, multi-protocols, and multi-technology environment.

### Prototype testbed to implement SDN features

The 6G-based SDN's various features, such as a controller node provides responsibility for collecting routing information and making routing decisions centrally. The SDN data plane will be the hypervisor (compute node) in the prototype model and the Software-Defined Networking functionality in a private cloud built using OpenStack.

### Energy-efficient edge computing model

Develop an energy-efficient edge computing model using SDN, NFV, and blockchain technology to mitigate or prevent end-to-end delay, network bandwidth, other network vulnerabilities, and various passive and active network attacks.

# 6  Conclusion

We developed a model to estimate the computational time on the SDN-forensics architecture with cybersecurity through a 6G network that works in the blockchain-based IoT environment. Our study presented results to demonstrate the effectiveness of the model. We modeled the computation process and time estimation. We have concluded that the total time consumption on the 6G SDN-forensic network increases with the time consumption of controllers and the time consumption on switches regardless of the number of IoT devices used. On the other hand, time consumption is higher in controllers than switches, and time consumption increment is higher in controllers than switches. Therefore, our study enhanced the understanding of the computational time on 6G-based SDN-forensic architecture in a blockchain-based IoT environment. However, the current study was not specifically designed to evaluate the computational time estimation on SDN for each application. The current findings substantially add to our understanding of signature validation's time duration, which is not a standard step in all applications. This chapter also provided a time estimation for achieving efficient results, with all the essential factors that influence the life expectancy on SDN-forensics architecture in the cybersecurity blockchain-based IoT environment.

# References

1. Aldiab M, Garcia-Palacios E, Crookes D, Sezer S (2008) Packet classification by multilevel cutting of the classification space: an algorithmic-architectural solution for IP packet classification in next generation networks. J Comput Syst Netw Commun 1–14. https://doi.org/10.1155/2008/603860
2. Pourvahab M, Ekbatanifard G (2019) An efficient forensics architecture in software-defined networking-IoT using blockchain technology. IEEE Access 7:99573–99588. https://doi.org/10.1109/access.2019.2930345
3. Pourvahab M, Ekbatanifard G (2019) Digital forensics architecture for evidence collection and provenance preservation in IaaS cloud environment using SDN and blockchain technology. IEEE Access 7:153349–153364. https://doi.org/10.1109/access.2019.2946978
4. Liu H, Zhang Y, Zheng S, Li Y (2019) Electric vehicle power trading mechanism based on blockchain and smart contract in V2G Network. IEEE Access 7:160546–160558. https://doi.org/10.1109/access.2019.2951057
5. Lu X, Shi L, Chen Z, Fan X, Guan Z, Du X, Guizani M (2019) Blockchain-based distributed energy trading in energy internet: an SDN Approach. IEEE Access 7:173817–173826. https://doi.org/10.1109/access.2019.2957211
6. Madan R, Mangipudi P (2018) Predicting Computer Network Traffic: A Time Series Forecasting Approach using DWT, ARIMA and RNN. 2018 Eleventh International Conference on Contemporary Computing (IC3). https://doi.org/10.1109/ic3.2018.8530608
7. Azi Laga S, Sarno R (2018) Time and Cost Optimization Using Goal Programming and Priority Scheduling. 2018 International Seminar on Application for Technology of Information and Communication. https://doi.org/10.1109/isemantic.2018.8549802
8. Aryan R, Brattensborg F, Yazidi A, Engelstad P (2019) Checking the openflow rule installation and operational verification. In: 2019 IEEE 44Th conference on local computer networks (LCN). https://doi.org/10.1109/lcn44214.2019.8990808.

9. Salehin K, Rojas-Cessa R, Lin C, Dong Z, Kijkanjanarat T (2015) Scheme to Measure Packet Processing Time of a Remote Host through Estimation of End-Link Capacity. IEEE Trans Comput 64(1):205–218. https://doi.org/10.1109/tc.2013.203

10. Yu H, Li K, Qi H (2019) An Active Controller Selection Scheme for Minimizing Packet-In Processing Latency in SDN. Security And Communication Networks 2019:1–11. https://doi. org/10.1155/2019/1949343

11. Ma L, Yang B (2019) Data Backup in Geo-Distributed Data Center Networks Under Time and Budget Constraints. 2019 International Conference on Networking and Network Applications (Nana). https://doi.org/10.1109/nana.2019.00036.

12. Haraty R, El-Kassar, A & Shebaro, B (2006) A Comparative Study of Elgamal Based Digital Signature Algorithms. 2006 World Automation Congress. https://doi.org/10.1109/wac.2006. 375953.

13. Halgamuge MN, Hettikankanamge S & Mohammed A (2020) Trust Model to Minimize the Influence of Malicious Attacks in Sharding Based Blockchain Networks. 2020 IEEE Third International Conference on Artificial Intelligence and Knowledge Engineering (AIKE). California, USA. https://doi.org/10.1109/aike48582.2020.00032.

14. Aiyar K, Halgamuge MN & Mohammad A (2021) Probability Distribution Model to Analyze the Trade-off between Scalability and Security of Sharding-Based Blockchain Networks. IEEE 18th Annual Consumer Communications & Networking Conference (CCNC). Las Vegas, USA. https://doi.org/10.1109/ccnc49032.2021.9369563.

15. Ekanayake O, Halgamuge M (2021) Lightweight Blockchain Framework using Enhanced Master-Slave Blockchain Paradigm: Fair Rewarding Mechanism using Reward Accuracy Model. Inf Process Manage 58(3):102523. https://doi.org/10.1016/j.ipm.2021.102523

16. Halgamuge MN, Mapatunage SP (2021) Fair Rewarding Mechanism for Sharding-based Blockchain Networks with Low-powered Devices in the Internet of Things. IEEE 16th conference on industrial electronics and applications (ICIEA'21) Chengdu, China

# Learning-Driven Nodes Profiling in 6G Wireless Networks: Vision, Challenges, Applications

Ranjana Sikarwar, Shashank Sheshar Singh, and Harish Kumar Shakya

**Abstract** The emergence of the new intelligent information world is increasing the responsibilities of the telecommunication industry to meet the unprecedented service-level needs for future applications like virtual reality (VR), 3D media, Internet of Everything (IoE), massive Machine-Type Communications (mMTC). To match this envisaged demand surge of service level and massive traffic volume, ultra-high data rates with reliable connectivity are required. 5G is deployed and used widely in many parts of the world. The application of automated systems with millions of sensors and chips embedded in them is popular in the healthcare industry, roads, space, oceans, homes, vehicles, and other environments to offer a smart and automated life. 6G wireless networks can show improvements over 5G with the increase in speed, reliability, and bandwidth availability. Also, it will enhance the quality of service (QoS) in key domain areas like healthcare, entertainment, smart cities, etc. 6G is expected to fulfill the crucial requirements for the high-end users and industry with a vision to focus upon key factors such as security and privacy. Blockchain is envisioned to address the security challenges in 6G and facilitate the functional standards.

**Keywords** 6G networks · Community detection · Complex profiling · Cognitive networks · Blockchain · Epidemic spreading

R. Sikarwar (✉)
Computer Science and Engineering, Amity University, Gwalior, Madhya Pradesh, India
e-mail: ranjana.sik@gmail.com

S. S. Singh
Computer Science and Engineering, Thapar Institute of Engineering and Technology, Patiala, India

H. K. Shakya
Computer Science and Engineering, Amity University, Gwalior, Madhya Pradesh, India
e-mail: hkshakya@gwa.amity.edu

© The Author(s), under exclusive license to Springer Nature Singapore Pte Ltd. 2022
M. Dutta Borah et al. (eds.), *AI and Blockchain Technology in 6G Wireless Network*,
Blockchain Technologies, https://doi.org/10.1007/978-981-19-2868-0_8

# 1 Introduction

The explosive growth in mobile traffic with the use of applications like [1–4] is expected to be 607 Exabytes/month by 2025. Also, the next generation is expected to be software-based, virtualized, and cloud-based systems connecting a large number of heterogeneous devices. These heterogeneous devices, for instance, IoT/IoE and other real-time data-intensive applications need to function at higher data rates with ultra-low latency. 5G networks will not be capable enough to fulfill the future emerging demands of the fully automated data-centric and intelligent network after 10 years [5]. 6G networks are envisioned to cater to this anticipated massive increase in data traffic. The software-based, virtualized, and cloud-based next-generation mobile networks offer many advantages in the future like agile and efficient management, micro-operator-based business models [6], and network orchestration (MANO) [7]. But these also exacerbate many challenges like security issues, soft-spectrum sharing, privacy, multiple access control, legitimate resource utilization, and many more. However, to establish trust in future networks blockchain a distributed ledger technology in general has been embraced widely by industry and many organizations [8–10]. The paramount number of advantages offered by indispensable blockchain technology such as decentralization, immutability, fewer processing delays, non-repudiation of the transaction, and transparency have proved it to be suitable for 6G wireless networks. The growing demand for bandwidth-intensive services and the use of high data-rate applications have forced the telecommunication industry the evolution of 6G. The next-generation 6G systems are designed by considering the requirements of higher data rates by the applications, gigantic antenna models [4], blockchain integration for security, artificial intelligence (AI), machine learning techniques, etc. Sixth-generation systems are expected to meet the demands of highly connected complex intelligent networks. Also, it will address the performance requirements of the innovative services dependent on increasing resources [11, 12]. The framework design process of 6G networks needs innovative approaches to change the systems from closed hierarchical patterns towards distributed networks. The networks design also needs to have self-organization and optimization capability for interaction between nodes (Table 1).

# 2 Background

As 5G wireless networks are the key enabler of information society 2020 [8], 6G network is the key enabling technology for intelligent information society 2030. The project of next-generation wireless local area networks IEEE803.11ax standard is also being developed. With the advent of massive MIMO technologies and millimeter-wave(mm-wave), 5G can provide end-users with high speed. Some of the projects undertaken for the development and promotion of 5G and 6G networks are listed in Table 2.

**Table 1** List of abbreviations

| AI | Artificial intelligence | Eurlcc | Extremely ultra-reliable and low-latency communications |
|---|---|---|---|
| 3D | Three dimensional | MBB | Mobile broadband |
| VR | Virtual reality | KPI | Key performance indicator |
| XR | Extended reality | Mr | Mixed reality |
| B5G | Beyond 5 G | MANO | Management and Network Orchestration |
| 4G | Fourth generation | MIMO | Multiple input multiple output |
| 5G | Fifth generation | Mm-wave | Millimeter-wave |
| 6G | Sixth generation | Thz | Terahertz |
| IoT | Internet of Things | SDN | Software-defined networking |
| IoE | Internet of Everything | NFV | Network function virtualization |
| QoS | Quality of Service | LIS | Large intelligent surfaces |
| IMT | International mobile telecommunications | HBF | Holographic beam forcing |
| VLC | Visible light communication | OAM | Orbital angular momentum |
| eMBB | Enhanced mobile broadband | ERLLC | Extremely reliable and low latency communication |
| URLLC | Ultra-reliable and low-latency communications | FEMBB | Further-enhanced mobile broadband |
| | | MMTC | Massive machine-type communication |
| umMTC | Ultra-massive machine-type communication | LDHMC | Long-distance and high-mobility communication |
| ELPC | Extremely low-power communication | CD | Community detection |
| EGT | Evolutionary game theory | M2M | Machine-to-machine |
| D2D | Device-to-device | UAV | Unmanned Aerial Vehicle |

# 3 Vision

The intelligent information society of 2030 is expected to integrate all functions including communications, fully digitalized intelligence-inspired systems with sensing, computing and caching capabilities with the help of 6G. 6G will connect everything with full wireless coverage and integrate positioning, navigation control, etc. 6G will be an independent fully wireless network ecosystem and will use all senses (fingers, voice, neural waves) to communicate with smart intelligent devices. The key factors supporting 6G systems include artificial intelligence-driven communication, increased data security, and privacy, enhanced energy efficiency, tactile internet and low backhaul. Building upon the 5G vision, the 6G discussion is gradually taking momentum to empower a plethora of autonomous services for IoT devices,

**Table 2** Projects undertaken for 5G and 6G

| S. no | Project name | Research organization/Country | Target wireless networks |
|-------|--------------|-------------------------------|--------------------------|
| 1 | 6Genesis | Finland | 6G |
| 2 | Broadband communications and new networks | China | 2030 or beyond |
| 3 | 3rd Generation Partnership Project | – | 5G |
| 4 | TERRANOVA | European Commission's Horizon 2020 | Beyond-5G(B5G) |
| 5 | THz bands | U.S Federal Communications Commission (FCC) | 6G |
| 6 | ITU-T Focus Group Technologies establishment | International Telecommunication Union (ITU) | Network 2030 |

super-smart cities, driverless cars and many more. The concept of flying taxis in large cities is one of the future visions which is already operating in countries like Dubai. 6G will be artificial intelligence (AI) inspired at almost all levels.

# 4 Requirements and Applications

6G will fill the gap between increasing market demands and laggings created by 5G in various issues of increasing demands of several connected devices, higher data rates, global connectivity. reliability, latency, throughput, energy efficiency, hardware complexity, deployment costs, reducing the energy consumption, battery-free IoT devices, integration of machine learning capabilities (Table 3).

The service requirements of 6G includes massive machine-type communications (mMTC), ultra-high-speed and low-latency communications (uHSLLC), ubiquitous mobile-ultra-broadband (uMUB), ultra-high density (uHDD). Table 4 summarizes the network characteristics of 5G and 6G. As compared to 5G, 6G systems are expected to provide 1000 times higher concurrent wireless connectivity. The evaluation of 6G wireless networks is based on some key performance indicators (KPIs) including peak data rates, user-experienced data rates, area traffic capacity (or space traffic capacity), latency connectivity density and mobility, and new KPIs other than those of used in 5G as shown in Table 3.

**Table 3** KPI comparison of 4G, 5G, and 6G

| Key performance indicators | 4G | 5G | 6G |
|---|---|---|---|
| Experienced data rate | 10 Mb/s | 0.1 Gb/s | 1 Gb/s |
| Peak data rate | 100 Mb/s | 20 Gb/s | $\geq$ 1 Tb/s |
| Latency | 10 ms | 1 ms | 10-100 $\mu$s |
| Spectrum efficiency | 1x | 3 times of 4G | 5-10times of 5G |
| Mobility | 350 km/h | 500 km/h | $\geq$ 1,000 km/h |
| Network energy efficiency | 1x | 10–100 times of 4G | 10–100 times of 5G |
| Area traffic capacity | 0.1 Mb/s/m$^2$ | 10 Mb/s/m$^2$ | 1 Gb/s/m$^2$ |
| Connectivity density | 10$^5$ Devices/km$^2$ | 10$^6$ Devices/km$^2$ | 10$^7$ Devices/km$^2$ |
| Satellite integration | No | No | Fully |
| AI | No | Partial | Fully |
| THz communication | No | Very limited | Widely used |
| Service level | Video | AR, VR | Tactile |
| Architecture | MIMO | Massive MIMO | Intelligent surface |

**Table 4** Network characteristics of 5G and 6G

| | 5G | 6G |
|---|---|---|
| Network features | • Cloudization<br>• Softwarization<br>• Virtualization<br>• Network slicing | • Intelligence-inspired<br>• Softwarization<br>• Cloudization<br>• Virtualization<br>• Network slicing |
| Service participants | Human beings and things | Humans and world |
| Technologies | • Massive MIMO<br>• mm-Wave communications<br>• Cloud/Fog/Edge computing<br>• NOMA<br>• Ultra-dense networks<br>• SDN/NFV/Network slicing | • SM-MIMO<br>• THz communications<br>• AI/Machine learning<br>• Blockchain-based spectrum sharing<br>• OAM multiplexing<br>• Quantum communications and computing<br>• Laser and VLC<br>• LIS and HBF |
| Application types | • eMBB<br>• URLLC<br>• mMTC | • FeMBB<br>• ERLLC<br>• umMTC<br>• LDHMC<br>• ELPC |

## *4.1  Applications Areas of 6G Wireless Networks*

### 4.1.1  Integration of 6G and Internet of Everything (IOE)

6G is expected to integrate with IOE to promote massive machine type communications and various sensing devices intended for monitoring, sensing everything including things, data, people, etc. The applications connecting IOE range from smart cities, business, industrial and healthcare systems, etc.

### 4.1.2  6G Support for UAV-Based Applications

As 6G's support for higher data rates provides new avenues for UAV-based applications serving remote aircraft and control in defense, military, traffic monitoring providing higher data rates transmission in disaster-prone areas having no cellular setup.

### 4.1.3  Industry 5.0 Based Automation

6G is expected to be a key enabler for an Industry 5.0 based automation environment including robots and smart machines connecting everything in the industry wired or wirelessly. The combination of AI and 6G intends to boost the performance of working robots.

### 4.1.4  eHealthcare

With the advent of AI, holographic telepresence, edge computing, and 6G healthcare sector will be benefited. The introduction of 6G wireless networks will facilitate remote surgery, monitoring of healthcare systems remotely. The in-body sensors planted in human bodies are provisioned with battery-less communication technologies.

### 4.1.5  Extended Reality services

XR services including AR (Augmented Reality), VR, and MR are promotional characteristics of 6G communication systems. Wearable XR devices with enhanced features such as XR, high-definition images, and holograms promote the human-to-human and things communication.

# 5 Challenges in 6G Wireless Networks

As expounded by Behnaam et al. [10], the challenging factors of 6G and some of the perceptive issues pertinent to M2M communication [10] are discussed below:

## 5.1 The Unprecedented Traffic Demands

*Minimum Latency in Real-Tme Communications*

The future computing ecosystems will also comprise real-time communication, which needs to have high accuracy and nearly zero delays for error-free operation. The M2M and D2D communications need to work with minimal delays and high accuracy. Applications such as autonomous driving can function efficiently along with minimal delay in communication.

*Scalability*

Due to the massive connectivity of billions of devices in industrial ecosystems such as the introduction of massive Machine-Type Communications(mMTC), tailoring the design of a 6G systems framework for managing unprecedented traffic is a challenge.

*Synchronization in Industrial Application*

The time-critical application areas such as backbone systems including power distribution systems or vehicular networks need real-time synchronization for precise operation.

*Increased Throughput*

The 5G-based mission-critical systems consisting of parallel functioning of billions of devices need to handle the massive volume of transactions concurrently in real time.

## 5.2 Security Issues in Future 6G and Beyond Ecosystems

*Authentication and Access Control*

The data used in any kind of transaction or database needs to restrict its access to prevent unauthorized access. The basic requirement to maintain data security is to have any access control mechanism. The transactional authentication mechanisms may impede scalability in futuristic 6G systems.

*Confidentiality*

The future computing wireless ecosystems of smart devices such as IoT may be exposed or vulnerable to attacks or threats due to wireless connectivity. The conventional encryption techniques used such as symmetric encryption key may not prove to be much stronger in data privacy due to computational restrictions.

*Integrity*

The enormous amount of data generated by future systems needs to be accessed by legitimate users during data transmission.

## *5.3  Audit for Compliance Check and Security Standards*

The holders of the network ecosystem need to be checked for compliance and enhanced security standards can be checked by using a deep packet-level audit mechanism. This auditing of a large number of holders participating in the network ecosystem will be exigent in terms of security.

*Requirement for Higher Data Rates*

One of the most demanding requirements of future network ecosystems is higher data rates for real-time applications such as holographic communications, 3D ultra-video, virtual reality, etc.

*Device Resource Restrictions*

Device resource constraints may restrict the capabilities of encryption algorithms due to limited computational and storage capabilities resulting in divergence from the standard mechanisms.

## 6  Learning-Driven Communities in Complex Networks for 6G

Complex systems depend on the structure and nature of interconnectivity between their constituent elements describing the dynamics of the system. The new 6G technology has evolved into a radical revolution for the integration of new paradigms into the interacting entities.

The upcoming 6G services are expected to design a network of heterogeneous nodes with different inner features, which can dynamically interact with each other unpredictably and randomly without any prior planning. The interactive elements of the network include individuals, their interests and behavior, wearable or handheld devices, etc., which form heterogeneous nodes forming a complex socio-technical ecosystem. The nodes of multiplex social networks consist of heterogeneous smart devices, mobiles, devices, etc. The entities of IoT subnetworks along with other smart devices are also represented as nodes of a multiplex social network. The multiplex dimension provides a suitable representation for the structural analysis of complex systems. Also, the 6G networks can be studied as a collection of multidimensional subnetworks comprising elements interacting in different ways. The transition from a closed network structure to an open distributed system or dynamic topology including vast heterogeneity necessitates the node-level intelligence. A node in a network is responsible for content diffusion, learning, and computation. Thus, a node profiling process in complex networks serves the above purpose [13]. The complex network's approach for node profiling integrates the phenomenon of using dynamics of diffusion, application of evolutionary game theory approach, applying models of epidemic spreading, hierarchical organization of the network into mesoscopic structures called communities.

*Multiplex Social Network*

The hidden dynamics of the complex systems can be studied by representing the network as a multidimensional network structure. This multidimensional structure can unveil the interesting topological characteristics and interaction between the elements of the complex networks.

Multidimensional representation of the network helps in investigating the important structural properties of the elements of the network and their interdependence as compared to the single graph representation in form of nodes and links connecting them [14, 14]. The multidimensional network structure can retain the important knowledge of structural complexity and connectivity, which may be lost in single graph representation. The important details and phenomena of real-world networks may be lost or missed by studying the network as a single layer only. Multilayer networks are the study of network structures at different layers of connectivity [16]. In this type of network, nodes have different layers of connectivity, which is a generalization of graph theory in networks. Complex connectivity of the networks is clearly described in multilayer networks as various channels of interaction are embedded in different layers. Multilayer networks also known as multiplex networks constitute many real-world networks like social networks, brain networks, biological networks, technological networks, social contagions [17], etc. The multiplex nature of social networks adds dimension to the study. The complexity of connections can be analyzed deeply by characterizing interactions based on cost, weight, and intensity in different layers [18].

## 6.1   Community Detection in Complex Networks

The rich structure of the multiplex network can be seen in the communities in complex networks [19]. Several algorithms for CD are proposed by many researchers. The study of networks on a mesoscopic scale, i.e., the intermediary structures called communities or clusters are analyzed to find the interesting patterns or hidden organization of the nodes [20]. The nodes of the network tend to organize or group themselves into clusters forming a nontrivial structure. The aim is to study the hidden organization of nodes and their interaction with other nodes' profiles, the layers, structure of multiplex networks, and characteristics of links.

## 6.2 Dynamics of Diffusion and Competition

The interplay of heterogeneous nodes and their ties in different layers of multiplex networks lead us to explore emerging phenomena like epidemic dynamics, super-diffusion, and cascading failures, etc. The diffusion dynamics are analogous to the spreading process of diseases [26], emotions, and misinformation. It can also be modeled as social contagion [21] in networks. The effect of diffusion processes evolving in the network is also an interesting part to be explored. The impact of content exposure and changes in dynamics of social behavior affect the diffusion process [27, 28]. The evolutionary dynamics of competition are analyzed using Evolutionary Game Theory (EGT) concept. EGT is an extension of the traditional Game Theory. Its applications can be seen in the study of genetics theory to find the frequency of a gene's appearance [29]. The study includes dynamism in strategies and the changes in the behavior of individuals interacting with each other in games. These theories of dynamism in strategies and gaming behavior of individuals help to predict the nature of real-world networks using mathematical models.

The purpose of EGT is to analyze the collective dynamics of cooperation and competition through social dilemmas. Different social dilemmas considered here are Prisoner's Dilemma (PD), Harmony Game (HG) [29, 29], Snowdrift Game (SD), etc. These social dilemmas are useful for studying the dynamics of cooperation. Figure 1 shows the node's profiling techniques. Profiling is detected in structural terms using the multidimensional structure of networks. The nodes are characterized based on diffusion and competition dynamics altogether. Finally, community detection techniques are used to analyze the role of nodes linked in the mesoscopic structure.

## 6.3 Structural Description of Nodes in 6G Network

A complex network is a collection of nodes with multiple links connecting and interacting with them. A multiplex social network $S_m$ is shown in Fig. 2 with different levels of interconnections among different layers representing a 6G subnetwork. A node indicates any basic units such as IoT devices, mobiles, laptops, gadgets or wearables, etc. The framework of the multiplex network is a population of N number

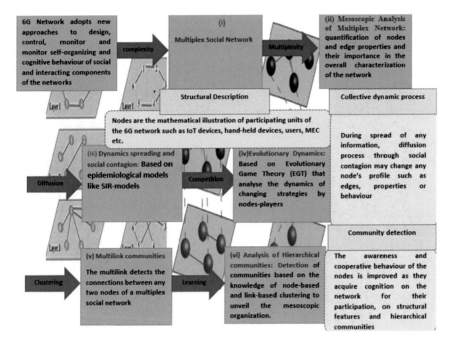

**Fig. 1** 6G network scenario modeling approach of heterogeneous nodes

of nodes and L number of layers as L1, L2, L3...L$_s$. Here, suppose s = 2 layers, which indicate L$_1$ as a real network layer [31] like…..., while the second layer is a virtual layer.

A multiplex network consists of a collection of graphs G. Each graph G can be defined as a set of vertices and edges (G = V, E). The number of nodes N is similar in each layer while the number of edges E varies with the layer [14]. Each network G$_L$ is represented in the form of an adjacency matrix $A^\alpha$ with elements $a_{ij} = 1$, if the nodes i and j are linked to each other on a larger L. The entry $a_{ij} = 0$ if there exists no connection between i and j. A weighted multiplex is represented by a weighted adjacency matrix $w^L$ with elements $w_{ij}$ [18]. In a weighted graph $a_{ij}^\alpha = w_{ij}^\alpha > 0$, there does exist an edge between i and j with a weight $w_{ij}^\alpha$, otherwise $a_{ij}^\alpha = 0$, $w_{ij}$ is a real positive number.

The basic features of profiling are defined below:

$w_{ij}$ denotes the weight of an edge between i and j,

$\deg_i^\alpha$ is the degree of a node i in layer L.

$O_{ij}^{[L1,L2]}$, a pair of nodes that can share connections in different layers or edge overlap.

$O_i$—overlapping degree.

The notations of the weighted multiplex network are as follows:

$O_i^w$—overlapping degree of a node with weights.

$O_{ij}^w$—weighted overlapping degree [18].

**Fig. 2** 6G network scenario
as multiplex social network

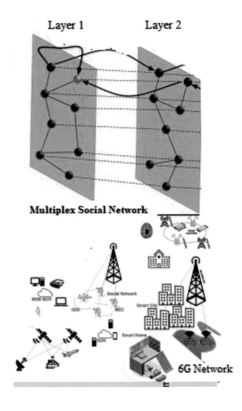

Social aspects can be reflected in structure profiling as shown in the equation below.

$$w_{ij}^{\alpha} = k_{ij}^{\alpha} |aw_i - aw_j| \tag{1}$$

$k_{ij}^{\alpha}$ is the tendency of similar nodes interacting with each other. Two nodes can share the same capacity of resources or are the same type of devices [28, 28, 28].

$aw$ is the node awareness coefficient.

The gap value between nodes i and j of node awareness $aw$ is represented by their difference.

We assume aw as the awareness level of a node i at a time T when we analyze the network activities. This awareness level for a node represents the acquired knowledge of the sub-network it is present in. The awareness level of a node i can be calculated as a function of attention level $atl_i$ and participation coefficient $\beta_i$.

$$aw_i|_{t=T} = aw_i|_{t=0} + \sum_T atl_{i|t-e} + P_i \tag{2}$$

The participation coefficient $P_i$ checks the awareness level by estimating the probability of collecting more knowledge from the information being exchanged across different layers [14, 14]. It adds the richness of the knowledge extracted from different layers of the multiplex network. The user-generated data is analyzed to compute the awareness. The attention level is also monitored to compute the awareness.

## 6.4  Study of Collective Dynamical Processes

The collective dynamical processes can be studied clearly from multiplex networks. To understand the complex interdependence of the networks, the multiplex social network provides a platform [32]. In this step, we consider the key features of evolving spreading processes. These key features are extracted from the analysis of quantifying estimators of two interdependent and co-evolving processes analogous to disease-spreading processes [21]. These spreading processes are thought of as "Composed-SIR" models. Composed-SIR models are derived from the classical SIR epidemic models [33]. Two interdependent spreading processes are referred to as the epidemic process of diffusion of awareness [21, 21]. For each node, awareness and heterogeneity are included. For describing the influence of the diffusion process and competition, heterogeneity and awareness are used for all nodes to favor the evolutionary dynamics. Each node represents the knowledge it has collected as awareness aw. This awareness is different for all nodes. The knowledge is acquired by the node from interest or participation in a collective phenomenon. Consequently, all the nodes will collectively join the diffusion process because the acquired awareness has an impact on the behavior of the node's functionality.

Initially, two coevolving epidemic processes are observed in the social multiplex network $S_m$. The first spreading process is according to the content shared based on common interest. The first spreading process can be expressed in terms of the reaction–diffusion equation taking into account the nature of shared content. Let the spreading processes be represented as $S^h I S^h$ or $S^h I R$.

$$S^h I S \Rightarrow S^h \xrightarrow{\beta_i^\alpha} I \xrightarrow{\mu} S^h \qquad (3)$$

$$S^h I R \Rightarrow S^h \xrightarrow{\beta_i^\alpha} I \xrightarrow{\mu} R \qquad (4)$$

where $S^h$ is defined as a heterogeneous susceptible state present in both models.

I—the state I means a condition when a node is infected.

R—recovered state.

The condition $S^h I S^h$ is when an interacting node is susceptible to joining the diffused process according to its interactions. In $S^h I R$-state, the last transition enters into the recovered state when a node stops participating in the diffusion process.

$\mu$ is the probability of call of the diffusion process, i is any node, each layer is denoted by $\alpha$, $\beta_i^\alpha$ is the diffusion rate of any node i at any layer $\alpha$ in the multiplex social network $S_m$ in which a node i in the layer $\alpha$ is prone to participation in the diffusion process. It is assumed that a node participating in the network is in the state of infection or informed.

**Community Detection Through the Multilink Structure of Nodes**

The entire multiplex network $S_m$ can also be observed as multilink [35]. The multilink structure of the network can describe the network's mesoscopic dimension. The mesoscopic dimensions of the multiplex network can be discussed through its multilink nature. All nodes collectively are linked through a multilink.

$(\vec{l}_{ij} = l_{ij}^{[L_1]}, l_{ij}^{[L_2]})$, where $L_1$ and L2 are the layers of the multiplex $S_m$.

If $\vec{l}_{ij} = 0$, two nodes have no interactions between them.

For community detection purposes, we consider the multidegree $\vec{k}_i^d$ of each node i of the population of nodes and the number of multilink $\vec{l}$ passing through node i.

A weighted aggregated network $\hat{N}^w$ is constructed using multilink $\vec{l}_{ij}$ and an adjacency matrix $A_{ij} = \phi(\sum a_{ij}^{|l|}$, $\phi(x)$ is a step function [35]. Communities are identified by looking at the multidegree for each node in the population and the number of multilink $\vec{l}_{ij}$ pointing on it.

Based on the above profiling steps, two community detection methods called node-based and link-based techniques are introduced for the mesoscopic network analysis scale. The resulting hierarchy consists of both the rich structure of the network and the collaborative dynamics in cognitive profiling. Since each node is a part of one or the other community, it dynamically adjusts to the changed behavior and social factors.

The definition of parameters used in community profiling are discussed below.

$\vec{l}_{ij}$—multilink between nodes i & j.

$\hat{N}^w$—weighted Aggregated network.

$A_{ij}$—Elements of $\hat{N}^w$.

$\vec{k}_i^d$—multidegree of node i.

# 7 Conclusions

This chapter highlights the various aspects of 6G wireless networks ranging from the vision for intelligent wireless networks for 2030, service requirements, the role of blockchain technology in enhanced security, intriguing challenges, and canvassed the role of node profiling in complex networks. The complex approach for 6G wireless networks with other tools and methods attempts to enable complex systems to be self-organized and cognitive networks. The introduction of the profiling approach in 6G networks helps nodes to achieve cognition ability to implement a fully intelligence-based distributed user-centric system. 6G in communication networks will change the future of the telecommunications network making them more resilient and self-organized. Complex networks reflect the collective behavior of the elements connected. The multiplex social network is studied in a mathematical form enabling different types of communication. The collective dynamic of diffusion is also studied which helps the communities in the network by increasing the cognitive capability of the nodes. Complex networks are fully described by the structural organization and connectivity of their elements. Complex networks describe the way their elements interact with each other collectively with their inner features. The future perspective of communication networks is expected to be changed by the involvement of 6G. Thus, complex systems with the incorporation of new methods and tools can design cognitive networks. These self-organized and resilient complex systems can serve many application domains like smart city networks and digital health care. Multiple types of interactions in a network can be studied through the complex profiling of heterogeneous nodes. The dynamics of diffusion are possible through social contagion and the approach of evolutionary game theory.

# References

1. Tariq F, Khandaker MR, Wong KK, Imran MA, Bennis M, Debbah M (2020) A speculative study on 6G.IEEE Wirel Commun 27(4):118–125
2. Piran MJ, Suh DY (2019) Learning-driven wireless communications, towards 6G. Proceedings of 2019 Int. Conf. Comput. Electron. Commun. Eng. iCCECE 2019:219–224
3. Fleetwood J (2017) Public health, ethics, and autonomous vehicles Am J Public Health 107(4):532–537
4. Saad W, Bennis M, Chen M (2020) A Vision of 6G wireless systems: applications, trends, technologies, and open research problems. IEEE Netw 34(3):134–142
5. Giordani M, Polese M, Mezzavilla M, Rangan S, Zorzi M (2020) Toward 6G Networks (2020): Use cases and technologies. IEEE Commun Mag 58(3):55–61
6. Yrjölä S (2019) Decentralized 6G Business Models. Proc. 6G Wirel. Summit, Levi, Finl., no. April, pp 5–7
7. Hewa T, Gur G, Kalla M, Ylianttila, Bracken A, Liyanage M (2020) The role of blockchain in 6G: Challenges, opportunities and research directions," 2nd 6G Wirel. Summit 2020 Gain Edge 6G Era, 6G SUMMIT 2020, pp. 4–8
8. Zhang Z et al (2019) 6G wireless networks: vision, requirements, architecture, and key technologies. IEEE Veh Technol Mag 14(3):28–41
9. Chowdhury MZ, Shahjalal M, Ahmed S, Jang YM (2020) 6G wireless communication systems: applications, requirements, technologies, challenges, and research directions. IEEE Open J Commun Soc 1(i):957–975
10. 6G Flagship, 6G Flagship, Key Drivers and Research challenges for Ubiquitous wireless Intelligence
11. Sergiou C, Lestas M, Antoniou P, Liaskos C, Pitsillides A (2020) Complex systems: a communication networks perspective towards 6G. IEEE Access 8: 89007–89030

12. Akyildiz IF, Kak A, Nie S (2020) 6G and Beyond: The Future of Wireless Communications Systems. IEEE Access 8:133995–134030
13. Scatá M, Attanasio B, Aiosa GV, La Corte A (2021) Cognitive profiling of nodes in 6g through multiplex social network and evolutionary collective dynamics. Futur. Internet 13(5)
14. Battiston F, Nicosia V, Latora V (2017) The new challenges of multiplex networks: Measures and models. Eur. Phys. J. Spec. Top. 226(3):401–416
15. De Domenico M, Nicosia V, Arenas A, Latora V (2015) Structural reducibility of multilayer networks. Nat Commun 6:1–9
16. Battiston F, Nicosia V, Chavez M, Latora V (2017) Multilayer motif analysis of brain networks. Chaos, 27(4)
17. Christakis NA, Fowler JH (2013) Social contagion theory: Examining dynamic social networks and humanbehavior. Stat Med 32(4):556–577
18. Menichetti G, Remondini D, Panzarasa P, Mondragón RJ, Bianconi G (2014) Weighted multiplex networks. PLoS ONE 9(6):6–13
19. Iacovacci J, Bianconi G (2018) S1 Appendix supplementary information on multilink communities of multiplex networks .Detailed description of the Multilink Community detection algorithm, pp 1–11
20. Iacovacci J, Wu Z, Bianconi G (2015) Mesoscopic structures reveal the network between the layers of multiplex data sets. Phys Rev E Stat Nonlinear Soft Matter Phys 92(4):1–14
21. Scatà M, Di Stefano A, La Corte A, Liò P (2018) Quantifying the propagation of distress and mental disorders in social networks. Sci Rep 8(1):1–12
22. Nicosia V, Skardal PS, Arenas A, Latora V (2017) Collective Phenomena Emerging from the Interactions between Dynamical Processes in Multiplex Networks. Phys Rev Lett 118(13):1–6
23. Santoro A, Nicosia V (2020) Algorithmic complexity of multiplex networks. Phys Rev X 10(2):21069
24. De Domenico M, Granell C, Porter MA, Arenas A (2016) The physics of spreading processes in multilayer networks. Nat Phys 12(10):901–906
25. Pastor-Satorras R, Castellano C, Van Mieghem P, Vespignani A (2015) Epidemic processes in complex networks. Rev Mod Phys 87(3):1–62
26. He X, Lin YR (2017) Measuring and monitoring collective attention during shocking events. EPJ Data Sci 6(1)
27. Scatá M, Attanasio B, Aiosa GV, La Corte A (2020) The dynamical interplay of collective attention, awareness and epidemics spreading in the multiplex social networks during COVID-19. IEEE Access 8:189203–189223
29. Newton J (2018) Evolutionary game theory: a renaissance.Games, 9(2)
30. Guo D, Fu M, Li H (2021) Cooperation in social dilemmas: a group game model with double-layer networks. Futur Internet 13(2):1–29
31. Magnani M, Micenkova B, Rossi L (2013) Combinatorial analysis of multiple networks, pp 1–17
32. Yang L, Yu Z, El-Meligy MA, El-Sherbeeny AM, Wu N (2020) On multiplexity-aware influence spread in social networks. IEEE Access, 8
33. Vespignani A (2012) Modelling dynamical processes in complex socio-technical systems. Nat Phys 8(1):32–39
34. Granell C, Gómez S, Arenas A (2014) Competing spreading processes on multiplex networks: awareness and epidemics. Phys Rev E Stat Nonlinear Soft Matter Phys 90(1):1–7
35. Mondragon RJ, Iacovacci J, Bianconi G (2018) Multilink communities of multiplex networks. PLoS ONE 13(3):18–22

# Investigating the Economic Impact of Organizational Adoption of AI-Enabled Blockchain in 6G-Based Supply Chain Management

**Ebru Gökalp and Elife Özer**

**Abstract** Artificial Intelligence (AI)-enabled blockchain technologies (BCT) in 6G-based Supply Chain Management (SCM) is a disruptive technological achievement that can affect business models and initiatives significantly. It provides various advantages such as developing innovative applications, lowering operation costs, increasing collaboration among supply chain partners, agility, transparency, accountability, decentralization, security, and interoperability. However, such novel technologies need to be adopted and used successfully by organizations to benefit the global economy. Despite their prospective advantages, even the most state-of-the-art technologies are susceptible to becoming obsolete if organizations do not adopt them. A theoretical understanding is necessary to define the underlying constructs encouraging or discouraging adoption and usage. Accordingly, this chapter investigates the constructs affecting the organizational adoption and usage of AI-enabled BCT in 6G-based SCM systems and the economic impact of this adoption. A systematic literature review method was followed to find out fundamental constructs in the literature. Then, a theoretical research model consisting of 17 constructs was developed by following the Technology-Organization-Environment Framework. After then, the economic impact of the organizational adoption of AI and BCT integration on the 6G-based SCM was analyzed.

**Keywords** AI-enabled blockchain · 6G · Supply chain management · Technology adoption · Technology-organization-environment framework

E. Gökalp (✉)
Department of Computer Engineering, Hacettepe University, 06800 Ankara, Turkey
e-mail: ebrugokalp@hacettepe.edu.tr; eg590@cam.ac.uk

Institute for Manufacturing, University of Cambridge, CB3 0FS Cambridge, England

E. Özer
Department of Management Information Systems, Adana Alparslan Türkeş Science and Technology University, 01250 Adana, Turkey
e-mail: efyilmaz@atu.edu.tr

# 1  Introduction

Supply Chain Management (SCM) includes managing business operations, consisting of logistics, purchasing, the flow of materials, and finance through an information and communication technology (ICT) infrastructure by integrating all peers in the supply chain including manufacturers, retailers, suppliers, wholesalers, and end-users into one single system. An effectively managed SCM system can significantly decrease a company's operating expenses, and increase profits. Therefore, it is indispensable to adopt and use this system to survive in the market and gain a competitive advantage [1].

The earlier definitions of SCM covered the flow of goods from supplier to manufacturer, distributor, and consumer, and then its scope was expanded further, covering the source of the earliest supply to end consumption [2]. Today, the SCM is a global phenomenon as supply chains are no longer confined to national boundaries. This means that the challenges of globalization are reflected in the SCM [3]. Interest in these challenges is growing both in the academic field and among practitioners. In this context, new solutions are sought to make transactions much more transparent, traceable, secure, effective, and faster at lower costs in a global supply chain. ICT has become an enabler to overcome these challenges by effectively integrating SCM actors and logistics processes. Moreover, further improvements in ICT will expedite more and longer distance trade [4]. At this point, the advances in blockchain technology (BCT), artificial intelligence (AI), wireless network technology, and a combination of these disruptive technologies provide a significant potential for innovative possibilities in the SCM.

BCT could provide convenient solutions such as decentralized management, robustness, immutable audit trail, privacy, and security [5, 6]. BCT applications in the SCM provide the physical tracking of supply chains, the security of SCM systems, and the use of smart contracts. Thus, organizations can enhance their supply chains' reliability, efficiency, and transparency and decrease overall operation and logistics costs by adopting BCT in their SCM systems [7]. Despite these promising benefits, a mere 9 percent of organizations have invested in BCT-based SCM systems [8]. Additionally, integrating AI into BCT can enhance BCT's underlying architecture and make AI more coherent and easier to understand. Even though the integration of AI and BCT remains partially discovered, both technologies are developing rapidly. Combining the key features of the technologies (e.g., AI's enabling analytics and decision-making from the collected big data as well as BCT's distributed, decentralized, immutable ledger used to store encrypted data) will bring innovative solutions. Moreover, the 6G technology infrastructure enables fast transactions, instant tracking, and speedy data sharing.

AI-enabled BCT in 6G-based SCM is likely to provide a disruptive effect on the global economy. It is predicted that the contribution of AI to the global economy will increase to approximately $16 trillion by 2030 [9]. Productivity growth in this contribution is expected to occupy a place of about $6.6 trillion, while the side effects of consumption are expected to be around $9.1 trillion. Similarly, BCT can boost the economy worldwide by $1.76 trillion in the same year. The value is primarily created by advancing tracking and tracing levels and enhancing trust. An AI-enabled BCT in 6G-based SCM can further improve the performance of SCM systems.

Considering these potential benefits of AI-enabled BCT in 6G-based SCM, the adoption of such systems constitutes a crucial source of competitive advantage. However, organizations do not adopt breakthrough technologies once their advantages are revealed. In other words, developing innovative technologies do not mean that they will be successfully accepted and utilized. Above all, technology adoption requires an initial awareness followed by prioritization of needs and examination of strategic alignments, benefits, difficulties, and barriers of the specific technology as a whole. Despite their challenges, organizations are more likely to adopt the emerging technologies if they are convinced of the potential economic benefits these technologies promise [10]. Therefore, a conceptual understanding is required to grasp the fundamental causes that encourage or discourage acceptance and utilization [10].

The present chapter analyses the constructs that impact the adoption and usage of AI-enabled BCT in 6G-based SCM through developing a research model formed on the Technology-Organization-Environment (TOE) framework [11]. Thus, adopting and using such an innovative system by the organization will be increased by adopting a solution implementation approach to transcend organizational obstacles; respectively, the economic impacts of adopting AI-enabled BCT in 6G-based SCM in organizations will be explored. To the best of the authors' knowledge, this chapter represents the first attempt in the literature. Thus, this study intends to fill this gap.

This study is formed of a few chapters. The following section gives comprehensive background information about the 6G technology, AI, BCT, SCM, and TEO framework. The research model development stages and methodology are explained in the following section. The proposed research model is explained in the fourth section and followed by the discussion section. Finally, some conclusions are drawn in the final section.

## 2 Background

This section delivers a review of the related studies, including 6G Technology, AI, BCT, SCM, and TOE framework.

### 2.1 6G Technology

6G technology refers to the sixth generation of wireless networks that promises high-speed internet connection and linking all network operators to one single core [12], although the 5G technology is not yet implemented at full scale worldwide, and there are still issues to be solved in 5G. The concept of beyond 5G (B5G) is dealing with these issues, enhancing the performance of 5G and thus, taking a further step to the next generation of wireless network, namely 6G.

Expectations are for 6G technology to be introduced by 2030 [13]. 6G will provide sophisticated capabilities to the networks. Device densities and Internet of Things (IoT) connectivity will be much denser than the 5G and up to 100 times faster. With 100 Gb per second peak data rates, latency, which can be reduced to microseconds, is one millisecond in the 5G. These main features have considerable potential to enable communication in almost real time.

Digitalization applications require high data and a great deal of connectivity for heavily embedded sensors and devices to catch tactile sensations and turn them into electronic data. 6G promises to enable a vast amount of data in nearly real time while ensuring security. The technologies underlying these features are expected to be AI and BCT; while AI improves the system performance, BCT can provide secure and flexible systems [14]. With these features, 6G can offer various advantages to business processes.

## 2.2   Artificial Intelligence (AI)

AI is a wide concept that embraces machines performing tasks that could be considered 'smart' such as reasoning, planning, and natural language understanding. Deep learning (DL) and machine learning (ML) are major subfields of AI. The essence of ML is giving machines the ability to learn without explicit programming. In other words, it enables machines to learn from the data. In addition, deep learning is a subfield of ML. DL algorithms are used for processing unstructured data (i.e., images, sounds, etc.).

AI techniques produce more accurate and meaningful results when applied to big data. Processing a large number of different data sources allows for evaluating interrelationships across various variables, leading to more intelligent outcomes for businesses. We are already familiar with AI applications such as recommendations engines, customer service, speech recognition, computer vision, and automated stock trading.

AI has a vital role in 6G autonomous networks. Deployment of AI techniques provides intelligence for wireless networks via learning and big data analytics.

## 2.3   Blockchain Technology (BCT)

In simple terms, a blockchain is an open database that holds a distributed ledger, typically in a peer-to-peer network [15]. The BCT is based on a cryptographic chain of blocks without a central repository or an authority, enabling a group of participants to record transactions in a collective ledger [16]. Each block connects to the previous block, creating a chain. The chaining process is executed via cryptographic assurance systems, and the data is immutable once published. This is enabled through the underlying consensus protocol, which is a procedure where all transactions are verified by all the nodes [17]. Accordingly, transaction history becomes precise and transparent across the node participants since each record forms an audit trail in the chains of blocks. Hence, this procedure increases the privacy and security of the wireless network [17].

Public, private, and consortium blockchains are the major BTC technology categories [18]. In a public BCT, participants can join without permission. This is the traditional BCT with no access limitation for verifying transactions, and creating a consensus block. Therefore, it is highly decentralized. A well-known example of a public blockchain is Bitcoin and Ethereum. Private blockchains, however, require inviting and permitting the selected participants by the network administrator for the specified transactions. Hyperledger can be an example of this type. Finally, consortium or hybrid blockchains are partly public and partly private. Pre-selected participants form a consortium, but only authorized participants can take part in the network. IBM Food Trust [19] platform is one of the examples of this type of BCT.

Businesses can benefit from BCT in various ways. Significant savings can be achieved in transaction costs and time in addition to the opportunities of enhanced transparency and security, advanced traceability, and increased efficiency. These benefits can create a tremendous shift in supply chain management systems. Several companies have already adopted BCT-based SCM systems and reported that the systems could reduce source control time from days or weeks to seconds [19].

## 2.4   Artificial Intelligence Enabled Blockchain

BCT and AI are advantageous technological developments for 6G-based SCM systems. The ability of BCT to form decentralized and secure resource sharing settings and the problem-solving features of AI is uncertain, time changing, and complex cases stand out in integrating these two technologies for promoting how wireless networks operate.

An example of an AI-enabled BCT network framework is provided by a recent study [20], which introduced a two-step consensus protocol via an outlier recognition method based on a machine learning algorithm. Tested in several IoT devices in a smart home network, the framework shows that a substantial increase in the fault tolerance of hyper ledger fabric can be achieved.

The exploitation of AI-enabled BCT can bring benefits in terms of [17]:

- ensuring secure and intelligent management,
- opportunities for flexible networking, and
- reliable and dynamic orchestration.

## 2.5   Supply Chain Management (SCM)

The SCM deals with material flow from suppliers to end-users in the production flow of a good or service. Suppliers, organizations, and customers are the leading players in a supply chain. Various upstream and downstream firms are involved in a supply chain, and raw materials, products, information, and financial transactions pass through the chain.

Integrating business processes with the leading supply chain participants is critical for an effective SCM [21]. ICT has become an enabler for effective integration within the SCM. It should be noted that the applications of emerging technology in the SCM will require comprehensive coverage, high-speed data transfer, low latency, multi-device connectivity, high security, and reliability [22]. At this point, 6G technologies and their impact on SCM are brought to the attention because of their promising potential, as described in the next subsection.

## 2.6   6G-Based Supply Chain Management

Mobile network technologies and their implications have been evolving at a fast pace. A recent study [1] examined the impact of 5G technology on SCM and concluded that 5G-based SCM offers significant performance improvement by enabling rapid response to changes via communication between the parties in the supply chain. Although the 5G-enabled SCM is in its early implication stage, the contribution of 6G is already a discussion matter with the advancements of network technologies and digital transformation.

6G can be considered as an enabler of the latest technologies, which are critical for the digital transformation of today's business, for instance, the IoT. In this context, 6G enhances real-time information flow capabilities, and thus, SCM performance is expected to be improved. A conceptual framework for AI-enabled BCT in 6G-based SCM is shown in Fig. 1.

## 2.7   TOE Framework

Adapting to developing ICT is vital in achieving a sustainable competitive advantage for today's businesses. Reducing operational costs, increasing revenues, satisfying customer needs, and improving business processes' effectiveness can be considered the most prominent advantages of effective ICT integration [23, 24]. The acceptance of innovations can be significantly affected by a firm's technological, organizational, and environmental contexts [25]. TOE is a widely accepted and utilized framework that provides valuable guidance to organizations in the acceptance of innovations. The framework explains the technological, organizational, and environmental contexts

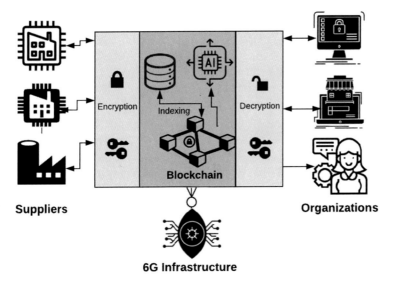

**Fig. 1** AI-enabled BCT in 6G-based SCM

as three major factors influencing adoption decisions. Although several models are studying the acceptance of innovative technologies, TOE presents a broader approach concerning the internal and external dynamics of the organizations [26]. Therefore, we utilized the TOE framework to determine the construct affecting the organizational adoption of AI-enabled BCT in 6G-based SCM in organizations to explore the economic impact of this novel technology.

## 3 Research Model Development

The Systematic Literature Review (SLR) technique suggested by Kitchenham [27] was utilized to determine the related studies and constructs affecting the organizational adoption of AI-enabled BCT in 6G-based SCM in the literature. Moreover, an expert panel consisting of senior academics and senior executive members working in technology management and SCM was constructed to evaluate if the determined constructs from the literature are adequate and appropriate. After reaching a consensus in the expert panel, the development of the research model was finalized.

## 3.1  Systematic Literature Review

SLR was performed for creating a proper database structure, including the recent works on the utilization and organizational adoption factors of the AI-enabled BCT in the 6G-based SCM domain. Research Questions (RQs) in line with these targets are determined as follows:

- RQ1: What are the existing studies related to the adoption of AI-enabled BCT in 6G-based SCM systems?
- RQ2: What are the fundamental constructs affecting the adoption of AI-enabled BCT in 6G-based SCM systems in the context of the organization?

After defining the aims of the SLR and corresponding RQs, the SLR was performed in September 2021. The details of the SLR operation are given in Table 1. As a result of entering the search query together with inclusion and exclusion criteria, 264 articles were identified at the first stage. After removing duplicate studies and evaluating the applicability of these studies according to the RQs based on their keywords, titles, and abstracts, 54 studies were found as relevant to the adoption of AI-enabled BCT in 6G-based SCM. After examining these 54 studies in detail, it was observed that 34 of them did not identify the critical elements for AI-enabled BCT adoption in 6G-based SCM, and therefore the remaining 20 studies were included as primary sources for examination.

**Table 1**  SLR Steps

| Steps | Explanation |
| --- | --- |
| Search Language | English |
| Search Query | The query for the search in title, abstract or keywords is ((ALL = (supply chain management)) AND ALL = ("blockchain" OR "artificial intelligence" OR "6G")) AND ALL = ("adoption" OR "acceptance")) |
| Database | Web of Science |
| Checking reference list | 264 articles were collected from databases |
| Inclusion and Exclusion Criteria | Inclusion: articles, book chapters, and conference proceedings Exclusion: Reviews, series, whitepapers, and meetings |
| Management of Results | Mendeley software is used |
| Selection of Primary Studies | After examining keywords, titles, and abstracts to understand the relevance of the studies concerning the RQs, 54 studies were found as mainly related to the adoption of the AI-enabled BCT in 6G-based SCM. The references to these studies were also reviewed |
| Study Quality Assessment | After assessing the quality of the primary studies, 34 studies were eliminated, and the rest of the 20 primary studies were deeply analyzed |

**Table 2** Constructs related to technology context from the literature

| Construct | [28] | [29] | [30] | [31] | [32] | [33] | [34] | [35] | [36] | [37] | [38] | [39] | [40] | [26] | [41] | [42] | [43] | [44] | [45] | [46] | # |
|---|---|---|---|---|---|---|---|---|---|---|---|---|---|---|---|---|---|---|---|---|---|
| Trust | | X | X | X | | X | X | X | X | X | | X | X | X | X | | | | | X | 13 |
| Relative advantage | | X | X | X | X | | | | | X | X | X | X | X | | | X | | X | | 11 |
| Complexity | | X | | | X | X | | | | | | | X | X | X | | X | X | X | X | 10 |
| Compatibility/ interoperability | X | X | | X | | X | | X | | | X | | X | X | X | | | | | | 9 |
| Scalability | X | | | | | | | X | | X | | | X | X | | | X | X | | X | 8 |
| Sustainability | X | | | | | | | X | | | | | | | X | | X | X | | | 5 |
| Performance expectancy | | | X | X | X | | | X | X | | | | | | | | | | | | 5 |
| Maturity | | | | | | | X | | | | | | | X | | | X | X | | X | 5 |
| Standardization | | | | | | X | | X | | | | | | X | X | | | | | | 4 |
| Triability and reversibility | | | | X | | | | | | | | | | | X | | | | | | 2 |
| Observability | | | | X | | | | | | | | | | | | | | | | | 1 |

**Table 3** Constructs related to organization context from the literature

| Construct | [28] | [29] | [30] | [31] | [32] | [33] | [34] | [35] | [36] | [37] | [38] | [39] | [40] | [26] | [41] | [42] | [43] | [44] | [45] | [46] | # |
|---|---|---|---|---|---|---|---|---|---|---|---|---|---|---|---|---|---|---|---|---|---|
| Organizations' technology readiness | X | X | X | X | | | X | | X | X | X | | X | X | X | | X | | | | 13 |
| Technical know-how | X | X | | X | | | X | X | | | | X | X | | X | | | X | | X | 10 |
| Top management support | | X | | | | | X | X | | | X | | X | X | X | | X | | | X | 9 |
| Financial resources | | | | | | X | X | | | | | | | X | X | | X | X | | X | 7 |
| Organization size | | | | | | | | | | | | | | X | | | X | | | | 2 |
| Perception vendor lock-in | | | | | | | | | | | | | | | X | | | | | | 1 |

**Table 4** Constructs related to environment context from the literature

| Construct | [28] | [29] | [30] | [31] | [32] | [33] | [34] | [35] | [36] | [37] | [38] | [39] | [40] | [26] | [41] | [42] | [43] | [44] | [45] | [46] | # |
|---|---|---|---|---|---|---|---|---|---|---|---|---|---|---|---|---|---|---|---|---|---|
| Government policy & regulations & support | X | | | X | | X | | X | X | X | | X | X | X | X | | X | X | X | X | 14 |
| Trading partner attitude (pressure or resistance) | | | X | X | X | X | | | | | X | | X | X | X | X | X | X | X | X | 13 |
| Inter-organizational trust | X | X | X | | | X | X | X | X | X | | | X | X | X | X | X | | | | 13 |
| Market dynamics | | | X | | | | | | | X | | | X | X | X | X | X | | X | X | 9 |

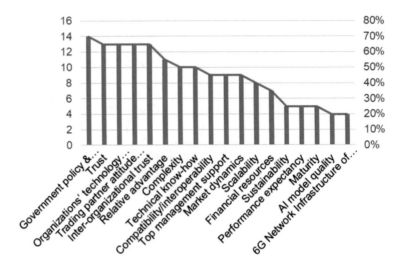

**Fig. 2** The frequencies of the constructs

As a result of the SLR, it was observed that 20 primary studies [26–46] focus on the significant adoption constructs in AI-enabled BCT in 6G-based SCM systems. The number of studies focusing on analyzing constructs of adoption and usage of AI-enabled BCT in 6G-based SCM is limited; 18 are related to the acceptance of BCT in SCM, 2 of them are related to the AI adoption in SCM. Therefore, it was concluded that there is no study exploring the adoption of AI-enabled BCT in 6G-based SCM in the context of the organization. Thus, the aim of this study is to bridge this identified gap by developing a research model including the constructs of the organization's adoption of AI-enabled BCT in 6G-based SCM. Constructs related to Technology, Organization and Envrionment contexts from the literature are given in Tables 2, 3 and 4, respectively. As seen in Fig. 2, we analyzed the frequencies of the construct. The highest frequent construct is government policy and regulation, which is mentioned in 14 of 20 primary studies.

## 3.2 The Expert Panel

The 20 primary studies obtained as a result of the SLR were used as the basis for developing the research model. However, because the number of existing studies is limited and the domain is still at the infancy stage, the constructs derived from the literature are not enough to develop a comprehensive research model. Thus, to get contribution from the industrial experts, an expert panel consisting of two senior academicians, one chief technology officer (CTO), and one senior employee in the SCM domain was constructed to contribute to the research model. Two consecutive sessions were conducted. Constructs from the SLR were analyzed in the first session; additional constructs offered by the experts were discussed in the second session, conflicts were resolved, and a consensus was reached on the research model, including the constructs of organizational adoption AI-enabled BCT in 6G-based SCM.

# 4  Proposed Research Model

The research model developed as a result of SLR is given in Table 5. A total of 17 determinants were classified under technological, organizational, and environmental contexts, as offered by the TOE framework [11].

**Table 5**  Proposed research model

| Context | Construct | Reference |
|---|---|---|
| Technology | Trust | [26, 29–31, 33–37, 39–41, 46] |
| | Relative advantage | [26, 29–32, 36–38, 40, 43, 45] |
| | Complexity | [26, 29, 32, 33, 36, 40, 41, 43–45] |
| | Compatibility/interoperability | [26, 28, 29, 31, 33, 35, 38, 41, 43] |
| | Scalability | [26, 28, 35, 37, 41, 44, 46] |
| | Sustainability | [28, 35, 43, 44, 46] |
| | Performance expectancy | [30–32, 35, 36] |
| | Maturity (Lack of Successful examples) | [34, 41, 43, 44, 46] |
| | AI model quality | Expert Opinion |
| Organization | Organizations' technology readiness | [26, 28–31, 34, 36–38, 40, 41, 43] |
| | Technical know-how | [28, 29, 31, 34, 35, 39–41, 44, 46] |
| | Top management support | [26, 29, 34, 35, 38, 40, 41, 43, 46] |
| | Financial resources | [26, 33, 41, 43, 44, 46] |
| Environment | Government policy and regulations and support | [26, 28, 31, 33, 35–37, 39–41, 43, 44, 46] |
| | Trading partner attitude (pressure or resistance) | [26, 30–33, 38, 41–46] |
| | Interorganizational trust | [26, 28–30, 33–37, 41–43, 45, 46] |
| | Market dynamics | [26, 30, 37, 41–43, 45, 46] |
| | 6G Network Infrastructure of the Country | Expert Opinion |

## 4.1  Technological Context

The technological context, comprising technology-related components, consists of Trust, Relative advantage, Complexity, Compatibility/interoperability, Scalability, Sustainability, Performance expectancy, Maturity (Lack of Successful examples), and AI model quality in the proposed research model.

### 4.1.1  Trust

Trust is considered as *"the perception of trustfulness in terms of system performance, availability of the system and possible risks involved"* from a technology adoption point of view [47]. Uncertainties in technology adoption create vulnerability for the adopter; therefore, it is essential to adopt a new technology [48] as security and privacy are the most substantial aspects of BCT. Some studies [49–51] remarked that these constructs would not impact AI-enabled BCT in 6G-based SCM adoption; nonetheless, [52] highlights it as a notable factor. Trust is found to be an essential variable and was added to the model.

### 4.1.2  Relative Advantage

The relative advantage of innovation is described as *"the degree to which the innovation is perceived to be more advantageous than the current one"* [53]. It is positively associated with the adoption of innovations [54]. Moore and Benbasat [55] specify the relative advantage's building blocks as a perceived advantage, perceived usefulness, and performance expectation in the context. Perceived usefulness refers to the user's perception of the new application that will likely improve the performance in the organizational context [56–58]. AI-enabled BCT in 6G-based SCM systems increases transparency and, therefore, increases employees' performance and productivity [59]. Accordingly, the relative advantage is included in the research model.

### 4.1.3  Complexity

Complexity can be described as *"the degree of the perceived difficulty in terms of understanding and using an innovation"* [53]. If users spend less effort using and learning a BCT, they will be much more motivated to use the system [26]. Thus, complexity is negatively related to adoption. AI-enabled BCT's decentralized structure has the feature to minimize this complexity by eliminating the transaction verification process complexity that may occur using a centralized tool. Therefore, the complexity construct is included in the proposed model, considering it will affect the organizational acceptance of AI-enabled BCT-based SCM.

### 4.1.4 Compatibility/Interoperability

Compatibility is "the degree to which an innovation is perceived as being consistent with the existing values, past experiences, and needs of potential adopters" [60]. Compatibility is found as a critical factor that affects the rate of the acceptance of any innovation [61, 62]. A new technology that does not conform to a system's current values and norms will not be adopted as quickly as a coherent innovation [60]. Therefore, organizations are more likely to adopt emerging technologies that are with their current infrastructure.

### 4.1.5 Scalability

Scalability covers extending BCT systems by adding new data. It is considered as the dissemination of change across different contexts in this study. Extending blockchain systems and assessing the costs of such extensions is difficult [51]. Makhdom et al. [63] suggest that increased BCT size can cause inefficiency and slow the transactions. As users enter data into the system, the BCT will expand to maintain the hash algorithm associated with the added data [26]. This expansion brings with it storage space and high processing power needs. Thus, it is considered a construct that can affect the adoption of AI-enabled BCT in 6G-based SCM.

### 4.1.6 Sustainability

Complex calculations and extracting blockchain in the network require a large amount of electricity [28]. In addition, with the increasing network load and new computational calculations, a personal computer may be insufficient to perform the required functions of a BCT-based system [28, 64, 65]. Therefore, organizations may also need to invest in high-functioning computers in the adoption process. In this context, high sustainability costs can be an obstacle to the adoption of an AI-enabled BCT in a 6G-based SCM; thus, it is added as a component in the proposed research model.

### 4.1.7 Performance Expectancy

Performance expectancy is described as "*the degree to which an individual believes that using the system will help him attain gains in job performance*" [66]. It was found as a construct influencing the adoption of BCT, SCM, and AI in different studies [30–32, 35, 36]. Thus, it is added as a construct in the proposed research model.

### 4.1.8   Maturity (Lack of Successful Examples)

Maturity can be defined as *"the richness of its resources and clear understanding of the technology"*. BCT, AI, and 6G networks are at their infancy stages, and they have not reached their maturation stage yet. Thus, there is a lack of successful examples in the market. Organizations and users can be afraid of adopting these new technologies because of uncertainties and possibilities of technical problems. Thus, it is thought to be an obstacle to adopting an AI-enabled BCT in a 6G-based SCM and added as a construct in the proposed research model.

### 4.1.9   AI Model Quality

The feature of AI models being able to learn over time is expected to enhance their performance in the long run. This feature is noteworthy, especially when the environment is dynamic in contexts such as supply chain and logistics where several partners, limited data, and data of questionable quality may be a subject[67]. Improving AI model quality is suggested to increase the potential to adopt AI-enabled BCT in SCM systems. Therefore, it is included in the proposed research model.

## 4.2   Organizational Context

The organizational context includes the tangible and intangible resources of an organization, and covers the components of technology readiness, technical know-how, top management support, and financial resources.

### 4.2.1   Organizations' Technology Readiness

Technology readiness is described as *"the people's propensity to embrace and use new technologies for accomplishing goals in home life or at work"* by Parasuraman [68]. Technology readiness is found as an essential aspect influencing a firm's adoption of emerging technologies [36]. The accurate IT infrastructure and human resources make it easier for firms to adopt new technology, whereas insecurity is an obstacle to technology readiness. Insecure potential users are likely to be skeptical about new ICT and may not even want to make an effort to see whether it will work for them or not. Nevertheless, optimism and innovativeness support technology readiness and drive the adoption process of new ICT such as BCT [45]. The fact that supply chain practitioners favor BCT will have a positive influence on the adoption of BCT-based SCM systems [45]. Therefore, it is considered a construct in the proposed research model.

### 4.2.2 Technical Know-How

Organizations with a technical background are suggested to be more likely to adopt BCT [28, 29, 31, 34, 35, 39–41, 44, 46]. The bounded technical know-how of using BCT constitutes a barrier to embracing this technology into the supply chain [69].

### 4.2.3 Top Management Support

Incorporating a new ICT into the organization is a strategic decision in itself. Accordingly, managers' knowledge and attitudes toward the new technology affect the adaptation decision. When the top management acknowledges the advantages of an AI-enabled BCT in 6G-based SCM, the requirements of the adaptation process will be fulfilled [51, 70]. For this reason, top management support is included in the research model, considering that it is an essential construct of the organizational adoption of AI-enabled BCT in SCM systems.

### 4.2.4 Financial Resources

Financial resources refer to the budget an organization allocates in transitioning to ICT innovations. The influence of the financial aspects on the adoption of BCT-based systems has already been highlighted in several studies [26, 41]. The greater the financial resources allocated to ICT innovation in organizations, the more inclined to initiate change, exert effort, and collaborate [61, 71]. As the implementation of BCT-based systems requires high initial investments [33, 41], financial resources are included in the proposed research model.

## 4.3  Environmental Context

Referring to the impacts from outside the organization, the environmental context consists of the Government policy and regulations and Support, Trading partner attitude (pressure or resistance), Interorganizational trust, Market dynamics, and 6G Network Infrastructure of the Country.

### 4.3.1 Government Policy and Regulations and Support

Government policy, regulations, and support encourage organizations to embrace emerging technologies and play an essential role in adopting technologies [41]. The absence or uncertainty of proper government regulations constitutes a barrier to adopting BCT [31, 33], and it discourages firms from considering the adoption first, while government encouragement policies are essential enablers for the adoption decisions [46]. Therefore, government policy, regulation, and support factors are included in the proposed research model, affecting organizations' adoption of BCT-based SCM technologies.

### 4.3.2   Trading Partner Attitude (Pressure or Resistance)

Trading partner pressure means the level of pressure confronted by the companies from their trading partners, and it is suggested as a critical factor in the adoption of new ICT [45, 72]. A blockchain-based SCM network needs collaboration with the members of the supply chain. In addition, synchronized operations among the supply chain partners using mutual systems are necessary [45]. In a scenario in which a dominant supply chain partner chooses to implement a BCT-based SCM system, other partners in the supply chain will be forced to deploy the system to continue the business partnership. Therefore, the pressure or resistance confronted by the trading partners is added to the proposed research model as a determinant that influences the adoption decisions of organizations on AI-enabled BCT in 6G-based SCM.

### 4.3.3   Interorganizational Trust

This construct is defined as "the willingness of a party to be vulnerable to the actions of another party based on the expectation that the other will perform a particular action important to the trustor, irrespective of the ability to monitor or control that other party" [73]. As stated by previous studies in the literature, it affects technology adoption [74, 75] and BCT adoption [30, 76]. Since BCT provides transparency and traceability, it improves cooperation through the SCM network [77]. Accordingly, it contributes to improving the level of trust in the SCM network [78]. Thus, it is included in the research model.

### 4.3.4   Market Dynamics

This construct covers the market ambiguity in regard to the acceptance of the technology. Proposing a valuable product to the supply chain can provide convincing customers to buy it [69]. Thus, market competition and demand uncertainty are also critical factors to adopt AI-enabled BCT in 6G-based SCM.

### 4.3.5 6G Network Infrastructure of the Country

As the last construct under the environmental context, network infrastructures of the countries are an essential aspect of adopting 6G-based SCM. Providing fast transmission across the peers in the supply chain requires an investment in 6G infrastructure by the governments of the countries located in the SCM system. Organizations located in a country without a proper 6G infrastructure could not successfully adopt AI-enabled BCT in 6G-based SCM.

## 5 Discussion

The future worth of AI in the global economy is estimated at an additional $13 trillion by the end of the third decade of the twenty-first century [79]. AI adoption across sectors (IT, energy, health care, smart cities, transport, etc.) and business management functions (SCM, customer service, health and safety, systems management, etc.) is based mainly on economic benefit expectations in terms of cost, time, performance, customer satisfaction, accuracy in decision-making and predictions [80].

In economic terms, the rise of global SCM has changed the dissemination of incomes across countries [4]. Peers on the supply chain can buy the products they need at a low cost with high trust and other market-related advantages. New collaborations and links between organizations can be established thanks quickly to AI-enabled BCT in 6G-based SCM because organizations trust the technology instead of their peers. Due to this technology, complex interconnections can be easily managed, solving problems arising from information asymmetry among partners and trust in financial transactions.

Considering all of these promising motivations, it can be asserted that this technology will drastically affect the global economy. However, it is crucial for organizations to take into consideration not only the technology but also the organizational and environmental aspects to successfully adopt this technology to overcome barriers to adopting and using it. The findings of the conducted SLR provide critical strategic implications to successfully design, develop, implement, and maintain AI-enabled BCT in 6G-based SCM systems. As part of the supply chain network, constructs defined under the context of the environment, such as government policy, regulations, and support, trading partner attitude (pressure or resistance), interorganizational trust, and market dynamics are critical. As an interorganizational ICT, SCM systems, one of the inter-organizational ICTS, are utilized for information transfer across organizations in the network. A shared SCM system must be used by partners, so the preference for transitioning to AI-enabled BCT in 6G-based SCMs is largely influenced by environmental variables. SCM systems providers should focus further on the leading companies in the supply chains as trading partner attitude (pressure or resistance) has a considerable significance on the adoption decision of AI-enabled BCT in 6G-based SCM systems.

# 6 Conclusion

A disruptive technology, AI-enabled BCT will be predicted to reshape the 6G-based SCM systems and business transactions across the countries by providing promising opportunities such as improving transparency, business analytics visualizations, high-speed transactions, immutability, real-time traceability, and efficiency. Correspondingly, it is estimated that the impact on the global economy of this innovative technology will be drastic. Despite these promising benefits, organizations have not adopted this technology yet. It is still in an infancy stage. To take advantage of this technology, organizations need to remove the barriers before adopting and using it successfully. Thus, this study aims to provide novel and reliable insights for ICT service providers to devise their SCM systems and companies to overcome obstacles and resistance to adoption and use.

This chapter makes substantial contributions by conducting an SLR to find out constructs affecting the adoption and usage of AI-enabled BCT in 6G-based SCM systems. Subsequent to the SLR, it became more apparent that no study aims to unify the adoption of AI, BCT, 6G-based network, and SCM. This chapter addresses this gap in the research literature. The other contribution of this study is to develop a theoretically grounded research model by following a well-known ICT adoption theory by identifying the primary contexts and related constructs to contribute to the adoption of this technology.

As a future investigation, a quantitative research method is planned to determine the most critical constructs affecting the adoption of AI-enabled BCT in 6G-based SCM systems in the context of the organization because one major limitation of the present study is the lack of quantitative data collection from responders with various backgrounds.

# References

1. Taboada I, Shee H (2021) Understanding 5G technology for future supply chain management. Int J Logist Res Appl 24. https://doi.org/10.1080/13675567.2020.1762850
2. Cooper MC, Lambert DM, Pagh JD (1997) Supply chain management: more than a new name for logistics. Int J Logist Manag 8 . https://doi.org/10.1108/09574099710805556
3. Meixell MJ, Gargeya, V.B.: Global supply chain design: A literature review and critique. Transp. Res. Part E Logist. Transp. Rev. 41. https://doi.org/10.1016/j.tre.2005.06.003
4. Park, A., Nayyar, G., Low, P.: Supply Chain Perspectives and Issues. (2013). https://doi.org/10.30875/a81e684f-en.
5. Gökalp, E., Gökalp, M.O., Çoban, S., Eren, P.E.: Analysing opportunities and challenges of integrated blockchain technologies in healthcare. In: Lecture Notes in Business Information Processing. pp. 174–183. Springer (2018). https://doi.org/10.1007/978-3-030-00060-8_13.

6. Gokalp, E., Coban, S., Gokalp, M.O.: Acceptance of Blockchain Based Supply Chain Management System: Research Model Proposal. In: 1st International Informatics and Software Engineering Conference: Innovative Technologies for Digital Transformation, IISEC 2019 - Proceedings. pp. 1–6. IEEE (2019). https://doi.org/10.1109/UBMYK48245.2019.8965502.
7. Francisco K, Swanson D (2018) The supply chain has no clothes: technology adoption of blockchain for supply chain transparency. Logistics 2:2
8. Caradonna, O.: Gartner supply chain executive conference.
9. PwC: Total economic impact of AI in the period to 2030.
10. Grublješič T, Jaklič J (2015) Business Intelligence Acceptance: The Prominence of Organizational Factors. Inf Syst Manag 32:299–315. https://doi.org/10.1080/10580530.2015.108 0000
11. Tornatzky LG, Fleischer M, Chakrabarti AK (1990) Processes of technological innovation. Lexington Lexington books. https://doi.org/10.1080/1550428X.2010.490902
12. PanimalarS, A.M. et al: 6G Technology. Int. Res. J. Eng. Technol. (2017).
13. Nayak, S., Patgiri, R.: 6G Communication: Envisioning the Key Issues and Challenges. EAI Endorsed Trans. Internet Things. 6, (2021). https://doi.org/10.4108/eai.11-11-2020.166959.
14. Gui, G., Liu, M., Tang, F., Kato, N., Adachi, F.: 6G: Opening New Horizons for Integration of Comfort, Security, and Intelligence. IEEE Wirel. Commun. 27, (2020). https://doi.org/10.1109/MWC.001.1900516.
15. Chen, Y.: Blockchain tokens and the potential democratization of entrepreneurship and innovation. Bus. Horiz. 61, (2018). https://doi.org/10.1016/j.bushor.2018.03.006.
16. Yaga, D., Mell, P., Roby, N., Scarfone, K.: Blockchain Technology Overview - National Institute of Standards and Technology Internal Report 8202. NIST Interagency/Internal Rep. (2018).
17. Dai, Y., Xu, D., Maharjan, S., Chen, Z., He, Q., Zhang, Y.: Blockchain and Deep Reinforcement Learning Empowered Intelligent 5G beyond. IEEE Netw. 33, (2019). https://doi.org/10.1109/MNET.2019.1800376.
18. Niranjanamurthy, M., Nithya, B.N., Jagannatha, S.: Analysis of Blockchain technology: pros, cons and SWOT. Cluster Comput. 22, (2019). https://doi.org/10.1007/s10586-018-2387-5.
19. IBM: IBM launches its blockchain produce tracker, IBM Food Trust.
20. Salimitari, M., Joneidi, M., Chatterjee, M.: AI-enabled blockchain: An outlier-aware consensus protocol for blockchain-based iot networks. In: 2019 IEEE Global Communications Conference, GLOBECOM 2019 - Proceedings (2019). https://doi.org/10.1109/GLOBECOM38437.2019.9013824.
21. Lambert, D.M., Cooper, M.C.: Issues in supply chain management. Ind. Mark. Manag. 29, (2000). https://doi.org/10.1016/S0019-8501(99)00113-3.
22. Lu, Y.: Industry 4.0: A survey on technologies, applications and open research issues, (2017). https://doi.org/10.1016/j.jii.2017.04.005.
23. Chandio FH, Irani Z, Zeki AM, Shah A, Shah SC (2017) Online Banking Information Systems Acceptance: An Empirical Examination of System Characteristics and Web Security. Inf Syst Manag 34:50–64. https://doi.org/10.1080/10580530.2017.1254450
24. Dehgani R, Jafari Navimipour N (2019) The impact of information technology and communication systems on the agility of supply chain management systems. Kybernetes 48:2217–2236. https://doi.org/10.1108/K-10-2018-0532
25. Williams MD, Rana NP, Dwivedi YK (2012) A Bibliometric Analysis of Articles Citing the Unified Theory of Acceptance and Use of Technology. Presented at the. https://doi.org/10.1007/978-1-4419-6108-2_3
26. Gökalp E, Gökalp MO, Çoban S (2020) Blockchain-Based Supply Chain Management: Understanding the Determinants of Adoption in the Context of Organizations. Inf Syst Manag 00:1–22. https://doi.org/10.1080/10580530.2020.1812014
27. Kitchenham, B.: Procedures for performing systematic reviews. Keele, UK, Keele Univ. 33, 28 (2004)
28. Sahebi IG, Masoomi B, Ghorbani S (2020) Expert oriented approach for analyzing the blockchain adoption barriers in humanitarian supply chain. Technol Soc 63:101427. https://doi.org/10.1016/j.techsoc.2020.101427

29. Kamble, S.S., Gunasekaran, A., Kumar, V., Belhadi, A., Foropon, C.: A machine learning based approach for predicting blockchain adoption in supply Chain. Technol. Forecast. Soc. Change. 163, (2021). https://doi.org/10.1016/j.techfore.2020.120465.
30. Queiroz MM, Fosso Wamba S (2019) Blockchain adoption challenges in supply chain: An empirical investigation of the main drivers in India and the USA. Int J Inf Manage 46:70–82. https://doi.org/10.1016/j.ijinfomgt.2018.11.021
31. Upadhyay N (2020) Demystifying blockchain: A critical analysis of challenges, applications and opportunities. Int J Inf Manage 54:102120. https://doi.org/10.1016/j.ijinfomgt.2020. 102120
32. Sohn K, Kwon O (2020) Technology acceptance theories and factors influencing artificial Intelligence-based intelligent products. Telemat. Informatics. 47:1–14. https://doi.org/10.1016/ j.tele.2019.101324
33. Yadav VS, Singh AR, Raut RD, Govindarajan UH (2020) Blockchain technology adoption barriers in the Indian agricultural supply chain: an integrated approach. Resour Conserv Recycl 161:104877. https://doi.org/10.1016/j.resconrec.2020.104877
34. Kouhizadeh M, Saberi S, Sarkis J (2021) Blockchain technology and the sustainable supply chain: Theoretically exploring adoption barriers. Int J Prod Econ 231:107831. https://doi.org/ 10.1016/j.ijpe.2020.107831
35. Caldarelli G, Zardini A, Rossignoli C (2021) Blockchain adoption in the fashion sustainable supply chain: Pragmatically addressing barriers. J Organ Chang Manag 34:507–524. https:// doi.org/10.1108/JOCM-09-2020-0299
36. Wong LW, Tan GWH, Lee VH, Ooi KB, Sohal A (2020) Unearthing the determinants of Blockchain adoption in supply chain management. Int J Prod Res 58:2100–2123. https://doi. org/10.1080/00207543.2020.1730463
37. Dora M, Kumar A, Mangla SK, Pant A, Kamal MM (2021) Critical success factors influencing artificial intelligence adoption in food supply chains. Int J Prod Res. https://doi.org/10.1080/ 00207543.2021.1959665
38. Kumar Bhardwaj, A., Garg, A., Gajpal, Y.: Determinants of Blockchain Technology Adoption in Supply Chains by Small and Medium Enterprises (SMEs) in India. Math. Probl. Eng. 2021, (2021). https://doi.org/10.1155/2021/5537395.
39. Karuppiah K, Sankaranarayanan B, Ali SM (2021) A decision-aid model for evaluating challenges to blockchain adoption in supply chains. Int. J. Logist. Res. Appl. 1–22. https://doi.org/ 10.1080/13675567.2021.1947999
40. Mathivathanan D, Mathiyazhagan K, Rana NP, Khorana S, Dwivedi YK (2021) Barriers to the adoption of blockchain technology in business supply chains: a total interpretive structural modelling (TISM) approach. Int J Prod Res 59:3338–3359. https://doi.org/10.1080/00207543. 2020.1868597
41. Choi D, Chung CY, Seyha T, Young J (2020) Factors affecting organizations' resistance to the adoption of blockchain technology in supply networks. Sustain 12:1–37. https://doi.org/ 10.3390/su12218882
42. Ghode D, Yadav V, Jain R, Soni G (2020) Adoption of blockchain in supply chain: an analysis of influencing factors. J Enterp Inf Manag 33:437–456. https://doi.org/10.1108/JEIM-07-2019- 0186
43. Clohessy T, Treiblmaier H, Acton T, Rogers N (2020) Antecedents of blockchain adoption: An integrative framework. Strateg Chang 29:501–515. https://doi.org/10.1002/jsc.2360
44. Boutkhoum, O., Hanine, M., Nabil, M., Barakaz, F.E.L., Lee, E., Rustam, F., Ashraf, I.: Analysis and evaluation of barriers influencing blockchain implementation in moroccan sustainable supply chain management: An integrated IFAHP-DEMATEL framework. Mathematics. 9, (2021). https://doi.org/10.3390/math9141601.
45. Kamble S, Gunasekaran A, Arha H (2018) Understanding the Blockchain technology adoption in supply chains-Indian context. Int J Prod Res. https://doi.org/10.1080/00207543.2018.151 8610
46. Vafadarnikjoo, A., Badri Ahmadi, H., Liou, J.J.H., Botelho, T., Chalvatzis, K.: Analyzing blockchain adoption barriers in manufacturing supply chains by the neutrosophic analytic hierarchy process. Ann. Oper. Res. (2021). https://doi.org/10.1007/s10479-021-04048-6.

47. Şener, U., Gökalp, E., Eren, P.E.: Cloud-Based Enterprise Information Systems: Determinants of Adoption in the Context of Organizations. In: Communications in Computer and Information Science. pp. 53–66 (2016). https://doi.org/10.1007/978-3-319-46254-7_5.
48. Bahmanziari, T., Pearson, J.M., Crosby, L.: Is trust important in technology adoption? A policy capturing approach. J. Comput. Inf. Syst. 43, (2003). https://doi.org/10.1080/08874417.2003. 11647533.
49. Angelis, J., Ribeiro da Silva, E.: Blockchain adoption: A value driver perspective. Bus. Horiz. 62, 307–314 (2019). https://doi.org/10.1016/j.bushor.2018.12.001.
50. Yusof, H., Farhana Mior Badrul Munir, M., Zolkaply, Z., Li Jing, C., Yu Hao, C., Swee Ying, D., Seang Zheng, L., Yuh Seng, L., Kok Leong, T.: Behavioral Intention to Adopt Blockchain Technology: Viewpoint of the Banking Institutions in Malaysia. Int. J. Adv. Sci. Res. Manag. 3, (2018).
51. Wang H, Chen K, Xu D (2016) A maturity model for blockchain adoption. Financ. Innov. 2:12. https://doi.org/10.1186/s40854-016-0031-z
52. Wanitcharakkhakul, L., Rotchanakitumnuai, S.: Blockchain Technology Acceptance in Electronic Medical Record System. (2017).
53. Premkumar G, Ramamurthy K, Nilakanta S (1994) Implementation of electronic data interchange: An innovation diffusion perspective. J Manag Inf Syst 11:157–186. https://doi.org/10. 1080/07421222.1994.11518044
54. Choudhury, V., Karahanna, E.: The relative advantage of electronic channels: A multidimensional view. MIS Q. Manag. Inf. Syst. 32, (2008). https://doi.org/10.2307/25148833.
55. Moore GC, Benbasat I (1991) Development of an instrument to measure the perceptions of adopting an information technology innovation. Inf Syst Res 2:192–222. https://doi.org/10. 1287/isre.2.3.192
56. Davis FD (1989) Perceived usefulness, perceived ease of use, and user acceptance of information technology. MIS Q. Manag. Inf. Syst. 13:319–339. https://doi.org/10.2307/249008
57. Çaldağ, M.T., Gökalp, E., Alkış, N.: Analyzing Determinants of Open Government Based Technologies and Applications Adoption in the Context of Organizations. In: Proceedings of the International Conference on e-Learning, e-Business, Enterprise Information Systems, and e-Government (EEE). pp. 50–56. The Steering Committee of The World Congress in Computer Science, Computer …, Las Vegas, Nevada (2019).
58. Şener, U., Gökalp, E., Eren, P.E.: ClouDSS: A Decision Support System for Cloud Service Selection. In: Lecture Notes in Computer Science (including subseries Lecture Notes in Artificial Intelligence and Lecture Notes in Bioinformatics). pp. 249–261 (2017). https://doi.org/ 10.1007/978-3-319-68066-8_19.
59. Bartlett PA, Julien DM, Baines TS (2007) Improving supply chain performance through improved visibility. Int J Logist Manag 18:294–313. https://doi.org/10.1108/095740907108 16986
60. Rogers, E.M.: Diffusion of innovations. Simon and Schuster (2010).
61. Wang YM, Wang YS, Yang YF (2010) Understanding the determinants of RFID adoption in the manufacturing industry. Technol. Forecast. Soc. Change. 77:803–815. https://doi.org/10. 1016/j.techfore.2010.03.006
62. Cooper, R.B., Zmud, R.W.: Information Technology Implementation Research: A Technological Diffusion Approach. Manage. Sci. 36, (1990). https://doi.org/10.1287/mnsc.36. 2.123.
63. Makhdoom I, Abolhasan M, Abbas H, Ni W (2019) Blockchain's adoption in IoT: The challenges, and a way forward. J Netw Comput Appl 125:251–279. https://doi.org/10.1016/j.jnca. 2018.10.019
64. Vranken H (2017). Sustainability of bitcoin and blockchains. https://doi.org/10.1016/j.cosust. 2017.04.011
65. Çaldağ, M.T., Gökalp, E.: Exploring Critical Success Factors for Blockchain-based Intelligent Transportation Systems. Emerg. Sci. J. 4, 27–44 (2020). https://doi.org/10.28991/esj-2020-SP1-03.

66. Venkatesh V, Morris MG, Davis GB, Davis FD (2003) User acceptance of information technology: Toward a unified view. MIS Q. Manag. Inf. Syst. 27:425–478. https://doi.org/10.2307/30036540
67. Venkatesh V (2021) Adoption and use of AI tools: a research agenda grounded in UTAUT. Ann Oper Res. https://doi.org/10.1007/s10479-020-03918-9
68. Parasuraman, A.: Technology Readiness Index (Tri): A Multiple-Item Scale to Measure Readiness to Embrace New Technologies. J. Serv. Res. 2, (2000). https://doi.org/10.1177/109467050024001.
69. Saberi, S., Kouhizadeh, M., Sarkis, J., Shen, L.: Blockchain technology and its relationships to sustainable supply chain management. Int. J. Prod. Res. 57, (2019). https://doi.org/10.1080/00207543.2018.1533261.
70. Gangwar H, Date H, Ramaswamy R (2015) Understanding determinants of cloud computing adoption using an integrated TAM-TOE model. J Enterp Inf Manag 28:107–130. https://doi.org/10.1108/JEIM-08-2013-0065
71. Weiner BJ (2009) A theory of organizational readiness for change. Implement Sci 4:67. https://doi.org/10.1186/1748-5908-4-67
72. Low C, Chen Y, Wu M (2011) Understanding the determinants of cloud computing adoption. Ind Manag Data Syst 111:1006–1023. https://doi.org/10.1108/02635571111161262
73. Schoorman, F.D., Mayer, R.C., Davis, J.H.: An integrative model of organizational trust: Past, present, and future, (2007). https://doi.org/10.5465/AMR.2007.24348410.
74. Wu K, Zhao Y, Zhu Q, Tan X, Zheng H (2011) A meta-analysis of the impact of trust on technology acceptance model: Investigation of moderating influence of subject and context type. Int J Inf Manage 31:572–581. https://doi.org/10.1016/j.ijinfomgt.2011.03.004
75. Lin HF (2011) An empirical investigation of mobile banking adoption: The effect of innovation attributes and knowledge-based trust. Int J Inf Manage 31:252–260. https://doi.org/10.1016/j.ijinfomgt.2010.07.006
76. Sharma M, Gupta R, Acharya P (2020) Prioritizing the Critical Factors of Cloud Computing Adoption Using Multi-criteria Decision-making Techniques. Glob Bus Rev 21:142–161. https://doi.org/10.1177/0972150917741187
77. Lamming RC, Caldwell ND, Harrison DA, Phillips W (2001) Transparency in supply relationships: Concept and practice. J Supply Chain Manag 37:4–10. https://doi.org/10.1111/j.1745-493X.2001.tb00107.x
78. Reyna, A., Martín, C., Chen, J., Soler, E., Díaz, M.: On blockchain and its integration with IoT. Challenges and opportunities. Futur. Gener. Comput. Syst. 88, 173–190 (2018). https://doi.org/10.1016/j.future.2018.05.046.
79. Bughin, J., Seong, J., Manyika, J., Chui, M., Joshi, R.: Notes from the AI Frontier: Modeling the impact of AI on the world economy. Model. Glob. Econ. impact AI l McKinsey. (2018).
80. Cubric M (2020) Drivers, barriers and social considerations for AI adoption in business and management: a tertiary study. Technol Soc 62. https://doi.org/10.1016/j.techsoc.2020.101257.

# The Economic Impact of AI-Enabled Blockchain in 6G-Based Industry

Pawan Whig, Arun Velu, and Rahul Reddy Naddikatu

**Abstract** The growth of billions of data-intensive apps has worsened the problem of restricted data throughput in the 5G wireless communication network. To solve this challenge, researchers are creating state-of-the-art technologies to fulfill the increasing wireless needs of the 6G wireless communication standards. Although certain candidate technologies from the 5G standards will apply to 6G wireless networks, significant disruptive technologies that will ensure the requisite quality of physical experience to enable universal wireless connection are expected in 6G. To understand the vision and objectives of 6G, this book chapter will first offer a basic background on the history of several wireless communication technologies. Second, we present a broad overview of the suggested supporting technologies for 6G and new 6G applications such as multi-sensory, extended reality, digital replica, and blockchain. Also, this chapter includes a special section based on the economic impact of AI-enabled blockchain in the 6G based Industry. Following that, the technology-driven difficulties, social, psychological, health, and commercialization issues faced by implementing 6G, as well as potential solutions to these challenges, are thoroughly explored.

**Keywords** Blockchain · 6G · 5G · Technologies · Communication

P. Whig (✉)
Vivekananda Institute of Professional Studies, 110034, Pitam Pura, Delhi, India
e-mail: pawan.whig@vips.edu

A. Velu
Equifax Data Analytic, Atlanta, IT, USA
e-mail: gctarun@gmail.com

R. R. Naddikatu
University of Cumbersome, Cumbersome, USA
e-mail: rahulnadi40@gmail.com

# 1   Introduction

As energy and climate are continually addressing the global pending problem, modern renewable energy is a dominant solution. Clean energy is known as green energy that can be used to offset conventional energy because it releases fewer pollutants and is renewable. As a result, green energy infrastructure and platforms are increasingly being built, and there is an increasing need for renewable energy expansion.

In this chapter, we introduce a new exchange mechanism to unlock new energy resources, which could be used in the future, using the blockchain. A blockchain is an open leader, where all transactions online are registered and all transactions can be bound, submitted, or checked. Blockchain is a digitalized accounting records scheme, whereby all transactions are recorded in detail according to a series of cryptographic principles to avoid illicit interference [2].

Each node in a blockchain transforms into a prosumer, able to generate, sell, and buy electricity, as well as exchange it across a peer-to-peer network without the need for a central organization.

Trading is executed remotely without the involvement of a third party because it is built on smart contracts. Different exchange criteria can be set, but the simplest method of trade is when the sell and purchasing amounts match depending on a price set according to the overall energy level. This allows us to use energy more effectively by establishing a nontraditional active energy trading network.

At this period, as a number of power and transaction records stored in the blockchain network are being registered, we create private networks that participate only in the allowed nodes without fear of failure and attack to avoid attacking nodes buying and selling indiscreet resources.

# 2   Blockchain

The blockchain is a distributed data network infrastructure based on P2P. As the blocks are clustered into chains, the blockchain is named. In the blockchain per network, the user distributes and stores data from the history of transactions in block format. Since any node has a single private key and public key, the secret key and the hash function can be used to carry out a cryptographic signature on the transaction. Each node uses the public key to check that the transaction has indeed been signed by the digital signature subject [3]. The block that includes this transaction has a structure of "chains" that are constantly linked after a certain cycle during the time flow. Each user can review their transaction history by reading their logs since every user has a transaction history. Transactions not authenticated cannot then be stored in the block. Blockchain thus has three features: data confidentiality, safety, and decentralization [4].

Blockchain is suitable to provide the information because it provides information that can be stored in an immutable ledger immediately, shared and fully open, and which can be accessed by approved network members only. A network of blockchains will monitor orders, transfers, accounts, development, and much more. And because participants have a common understanding of the facts, all the transaction information is end-to-end and gives you increased interest and new efficiencies.

## 3  Aspects of AI With Blockchain

AI and blockchain are proven to be a potent combo, benefiting nearly every sector in which they are used. From agricultural supply chain and health sharing of information through media royalty and financial security, blockchain and AI technologies are teaming to improve everything. The combination of AI and Blockchain affects several elements, notably security—AI and blockchain will provide a second layer of defense against the internet.

Artificial Intelligence can successfully mine datasets to generate novel situations and uncover trends based on data behavior. Blockchain technology aids in the effective removal of flaws and fake data sets. AI-generated classifiers and patterns may be validated on a decentralized blockchain structure to ensure their legitimacy. This applies to any consumer-facing company, such as retail transactions. Data collected from clients using blockchain technology may be utilized to construct chatbots via AI.

### 3.1  How AI Can Be Included in Blockchain

The combination of AI with blockchain results in possibly the most dependable innovation judgment system in the world, one that is essentially trying to mess and gives solid insights and judgments. It has several advantages, including

- Improved corporate data models,
- Globalization verification systems,
- Innovative audits and conformance systems,
- Cleverer finance,
- Clear governance,
- Smart retail, and
- Smart prediction.

## 3.2 Technical Aspects

**Security**: Only with the incorporation of AI, blockchain technology become safer by providing better application deployments that are secure. An amazing example is AI systems that are progressively deciding if bank deposits are illegal and should be banned or probed.

**Efficiency**: Artificial Intelligence can assist in optimizing algorithms to minimize mining load, resulting in lower network latency and speedier transactions. AI enables blockchain technology to have a lower carbon impact. The expense imposed on miners, as well as the energy expended, would be decreased if AI devices replaced the labor done by mine workers. As the data on blockchains grows by the minute, artificial intelligence data pruning techniques may be used in the blockchain data for automatically pruning the data that is no longer needed for future usage. AI can even bring new decentralized learning methods, such as supervised learning, or new data-sharing mechanisms, which will greatly improve the system's efficiency.

**Trust**: One of blockchain's distinguishing features is its immutable records. When used in combination with AI, people will have surpassed the previous to track the system's thought process. This, in turn, increases the bots' trust in one another, enhancing machine-to-machine contact and enabling them to exchange data and coordinate large-scale decisions.

**Superior Management**: Whenever it breaks codes, human professionals improve with practice over time. A machine-learning-powered mining method can eliminate the need for human expertise because it can virtually completely perfect its abilities if given the necessary coaching knowledge. As a result, AI also aids in the better management of blockchain networks.

**Memory**: Blockchain networks are suitable for storing extremely sensitive, personal details, which may be added comfort and quality when efficiently handled with AI. Intelligent health systems that use medical scans and data to make the accurate diagnosis are an outstanding example of this.

## 4 Main Elements of Blockchain

The main elements of the blockchain are explained in the subsequent section.

## 4.1 Distributed Ledger

DLT is an asset transaction digital structure in which transactions and their data are registered simultaneously in several locations. DLT is the only way to register the transaction. In contrast to conventional databases, distributed ledgers don't have a central store or management function [5–7].

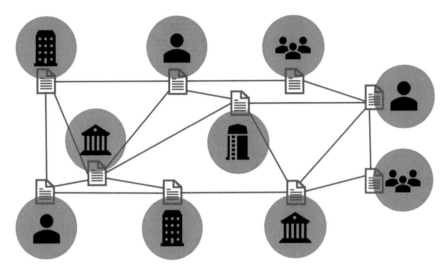

**Fig. 1** Distributed ledger technology

Each node processes and checks each item on a distributed ledger, generating a record of each item and agreement for its truthfulness. You may use a distributed directory for recording static records, such as a register and dynamic transactions.

This computing architecture reflects a major change in record-keeping by changing the collection and communication of records.

**Source of the Ledgers Books**

Ledgers–mostly transaction records and related figures—have remained on paper for thousands of years. They became digitized at the end of the twentieth century with the development of computers, while computerized books usually reflected what used to appear on paper. However, the validity of transactions reported in the records must be checked by a central authority in the period. For example, financial transfers need to be checked by banks [8–10].

The distributed directory and its unchangeable transaction log are accessible to all network members. This mutual directory records transactions only once, avoiding the repetition of effort characteristic of conventional corporate networks. Workflow in the distributed ledger is shown in Fig. 1.

## 4.2 Importance of DL

The distributed ledger technology will speed up transfers because it eliminates the need for a central agency or intermediary. Likewise, distributed booklets will reduce transaction costs [11–14].

Experts are also of the opinion that distributed ledger technological information is much safer when each network node has data and creates a mechanism that is harder to hack or target effectively.

Many often see a distributed block as a much more open method of managing records, as the content is being exchanged and thereby witnessed in a network [15].

The use of distributed ledger technologies in financial transactions was very important early. This is comprehensible because of the worldwide usage of blockchain bitcoin, although also showing that DLT may function [16]. Early innovation also became in this room in banks and other financial institutions.

However, in addition to financial purchases, DLT advocates claim that digital books can still be used in other fields, including government and industry. Experts think that automated leads will be used to collect taxes, pass land deeds, distribute social services, and also vote.

They also state that DLT can be used to process and implement legal documents and the like.

Some claim that people can use this technology to collect, monitor, and selectively distribute personal data when necessary. Use cases include medical records and company supply chains [17–20].

Proponents also believe digital booklets will help to monitor the rights and possession of literature, commodities, songs, movies, and more of intellectual property.

### 4.3    Futures of DL Systems

It is an ongoing question whether distributed leading innovations, such as blockchain, can revolutionize the way economies, organizations, and businesses operate. Academic and financial press articles asked whether distributed ledger systems as they currently exist are secure enough to be used widely. These modern modes of trade and security issues are not subject to any legislation.

## 5    Immutable Records

After the transaction is recorded in the public ledger, no user can alter or manipulate it. If an error occurs in a transaction log, the error must be reversed by adding a new transaction and all transactions are then available. Immutable record is represented as shown in Fig. 2.

The word "immutable" can almost always be found across hundreds of posts and discussions surrounding blockchain. Immutability—the ability to remain a constant, unchanging, and unalterable transaction background for a blockchain leader—is a

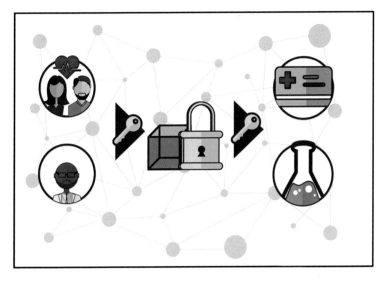

**Fig. 2** Representation of immutable record

definite attribute that blockchain evangelists emphasize as an important advantage. Immutability will turn the audit process into a quick, reliable, and cost-effective process to increase confidence and integrity in the use and sharing of data companies daily [21–25].

Trillions of dollars would be spent on strategies for cyber security that can secure our confidential data out of our sight. However, we seldom fight the domestic cyber protection fight: to ensure that our data is not compromised, substituted, or falsified by a corporation or its employees. In certain instances, we depend on methods such as private keys and user permissions to correct data. Yet, methodologically or mathematically, one cannot prove the knowledge is flawless in a typical program database. The next line of security is auditing [26–28].

The introduction of blockchain will provide a regular degree of confidence for data companies—immutability gives credibility. Blockchain allows customers to show that the information we use to have is not manipulated and at the same time the audit process can be turned into an effective, sensible, and economical operation.

**What is Immutability?**

A brief introduction to hazardous cryptography.

We need to consider cryptographical hatching before diving into blockchain immutability. The fundamentals are as follows:

A hash function takes the current data and outputs a "Checksum"—a number and letter string that serves as digital signatures.

The check amount is guaranteed to indicate your exact data entry—if any one byte differs from two files, the outputs are two strings after hacking. This can be associated with an avalanche effect [29].

SHA-2 (and its variants: Sha-256 is the most common in the blockchain world) is probably the most popular hashing algorithm. It was developed by the NSA.

Cryptography + Blockchain Hashing Process = Immutability.

Transaction validated by the blockchain network is time-marked and inserted in a "block" of records, encrypted by a hacking mechanism that links the hash in the previous block and integrates it into the chain as a sequential update [30–34].

Including metadata from the previous block, the hash output will be used for a new block. This connection during the hacking process renders the chain "unbreakable"—after validation, data cannot be manipulated or deleted so the subsequent blocks in the chain deny the change attempted. In other words, as data is manipulated, the blockchain breaks down and the cause can be found easily. This feature is not present in conventional databases, where details can be easily changed or erased [35].

At a certain time, the blockchain is simply a reference to the truth. For Bitcoin, these details include Bitcoin transmission detail. The picture below illustrates how as part of the header the check-summation of the transaction data has been applied to and becomes a checksum for the whole block [36–39].

## 6  Smart Contracts

To speed transactions, a set of rules—called a smart contract—is stored on the blockchain and executed automatically [40]. A smart contract can define conditions for corporate bond transfers; including terms for travel insurance to be paid and much more. A block diagram to represent the smart contract is given in Fig. 3.

Intelligent contracts are self-executing contracts bearing the terms of an arrangement between peers. The provisions of the arrangement shall be included in the code. The intelligent contract runs on the open network of the Ethereum blockchain [41–44]. The arrangements make capital, shares, property, or other commodities easier to trade. There are two commonly used smart contracts in Ethereum—Solidity and Serpent—programming languages. Solidity is an advanced programming language used on the Ethereum blockchain platform for integrating intelligent

**Fig. 3** Block diagram to represent Smart contact

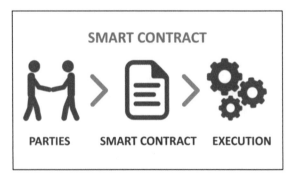

contracts. It allows developers of the blockchain to review the program, instead of compiling it at runtime [45].

Two parties traditionally use a reputable third party's expertise to carry out the arrangement as two parties enter into a deal. For millennia, it was handled this way. However, the implementation of intelligent contracts and their associated applications automate a difficult manual operation [46]. In 6G to make it more secure and highly reliable, Smart Contracts are needed in the future, this can be possible only with the application of blockchain. Hence, blockchain is an integral part of digital transactions to make the process highly secured. In future, smart Blockchain with the application of Artificial intelligence make 6G smart contracts more accurate, efficient, and secure.

### 6.1 Savings for Intermediaries, Automation, and Time

Taking days and even weeks to perform a conventional contract, the sheer number of lower and middle levels delays the process.

Smart contracts can be carried out on a computer under predefined environments for just minutes since they are automated and programmable. There is no involvement by third parties.

### 6.2 Safety

Contacts with standard contracts are issues about privacy and stability. Safety can be jeopardized at any time with too many intermediary parties involved. When using smart contracts, security is retained by encryption, the public key, and private keys. The data is almost difficult to change when held in a decentralized environment. Intelligent agreements are entered into remotely using private keys. The public key exchanged by the participating parties cannot be decoded [47].

### 6.3 Precision and Openness

Conditions and terms in an intelligent contract are predefined and pre-integrated. Once a condition has been reached, the sending happens and is registered automatically. If every transfer is part of a conventional contract, it is a manual procedure with workflows for consent. The groups concerned, peripheral bodies, and intermediaries traditionally dictate accountability. It's a device that's flawed. However, intelligent contracts are 100% open, and 24*7*365 are available online. Anyone will check and verify the transactions archived. Traditional contracts are impossible to archive since they are offline and dependent on paper [48].

## 6.4   Fare

Compared to smart contracts, traditional contracts are costly mainly because all the intermediaries must be paid for. There are no intermediaries for smart contracts, and only the exchange costs derive from the underlying smart contract network networks.

## 6.5   Benefits of Blockchain

What has to be changed: Activities also expend time to maintain redundant records and to validate third parties. Record-keeping systems can be vulnerable to cyber-assault and theft. Weak data verification can be slowed by limited clarity. And trade rates soared with the advent of IoT [49]. All this is slowing down the industry, draining the end – which suggests that we need a better direction. Some benefits of the blockchain are described in the following.

(a)   **Trust**

One should be sure that you receive reliable and timely data from blockchain as a part of a Network-Only, and that the private information can only be exchanged with network users to which you have specific access [50].

(b)   **Security**

Both network participants need a security consensus on data accuracy, and all authenticated transactions cannot be changed since they are forever registered. Nobody can uninstall a transaction, not even a system administrator.

(c)   **Sustainability**

The time-consuming record reconciliation is avoided with a public ledger exchanged by users of a network. And a series of laws—known as an intelligent contract—may be saved and exceeded instantly in a blockchain for pace transactions.

## 6.6   Blockchain Benefits for 6G

A new wireless network, 6G, will satisfy the needs of Internet of Everything applications and help usher in a data-intensive, intelligent society of the future, according to the company. Blockchain, according to researchers and industry experts, will help the 6G network's functioning requirements.

The most important technologies that will be the driving force for 6G are the terahertz (THz) band, AI, optical wireless communication (OWC), 3D networking, unmanned aerial vehicles (UAV), and wireless power transfer. The detailed discussion is out of the scope of this book chapter.

Sixth-generation communications (or 6G) will have a profound impact on the world by 2030. Even though many nations are currently working on 5G, several research organizations have already begun work on 6G. Cryptocurrency-based technologies might help overcome the existing constraints of the 5G network and enable sophisticated IoE applications for 6G. The potential of blockchain technology and its future possibilities will be discussed in this article.

Mobile networks of the sixth generation (also known as 6G) will provide a superior user experience by covering the whole air–space–sea–land (ASL) system. The 6G mobile traffic is expected to reach 607 Exabytes/month by 2025 and 5016 Exabytes/month by 2030, according to estimates.

When combined with artificial intelligence, the Internet of Things, and blockchain, 6G will be able to satisfy unprecedented service-level requirements, such as ultra-high data speeds and traffic volumes for applications. Virtual Reality, holographic communications, and huge machine-type communications are all possible uses.

To further appreciate how blockchain might aid 6G mobile networks, let's look at some of the perceptible and relevant problems that 6G faces.

The sixth-generation network is experiencing several challenges as it is being built. According to Behnam, Biral, and others, the following has been studied:

A large number of linked devices will be used in the future industrial ecosystems as a result of the commercial IoT with billions of gadgets. Adapting 6G systems to changing traffic conditions will be a difficulty, though.

Transmitting data in real time with the least amount of latency is essential for precise operations between devices and machines.

For industry-related essential applications, power distribution systems and vehicular networks must be integrated in a coordinated manner.

To manage the massive amount of real-time transactions created by future systems, the network infrastructure must be able to handle higher throughput.

Standards for validation and entry restrictions are necessary to prevent unauthorized modifications. The low-power IoT devices utilize lightweight cryptography methods that might expose the data.

As 6G networks get more complex, they will include an increasing number of interconnected devices, which will raise the risk of a Distributed Denial of Service assault.

An audit of tenant activity in the network system is essential. A vast number of renters will be audited, but implementing common security measures for the devices will be a challenge.

## 7  6G Advantages from the Blockchain:

This is how blockchain can solve the following problems:

- Securing distant resources and offloading computation as well as creating trust between edge servers and consumer devices.

- The enhancement of spectrum sharing security and the prevention of lease record manipulation.
- Assuring trust between cache requesters and suppliers when high-volume material is cached on user devices
- When sharing energy or resources via smart devices or trading platforms, ensuring confidence among market players.
- Optimized inference management by eliminating middlemen
- Eliminating the constraints of scaling by making enormous connections possible

However, scalability, Sybil attacks, and privacy leaks are still issues that need to be addressed before blockchain can be used to solve them. With the integration of blockchain with 6G, industries would benefit from transparency and information exchange through the use of 6G technology.

Assuring the market participants' confidence while exchanging energy or resources via smart devices or energy trading marketplaces is essential.

Optimized inference management and the elimination of middlemen,

eliminating restrictions on scalability by providing enormous connections.

However, scalability, Sybil attacks, and privacy leaks are yet to be solved. Transparency and information sharing would be aided by the incorporation of blockchain into 6G.

Blockchain's potential is limitless. While more rigorous research and ongoing innovation and technology breakthroughs will assure a bright future in the coming decades, aspiring blockchain technologists and business experts are needed to keep the needle rolling.

## 8   Applications of Blockchain with 6g

It is discussed how to categorize blockchain-industrial applications in 6G-enabled IoT. It included traditional areas of interest for various reasons such as smart manufacturing, supply chain management, the food industry, smart grid, health care, multimedia and digital rights management, agriculture, and the internet of vehicles and unmanned aerial vehicles as shown in Fig. 4.

Blockchain and 6G enhanced security and bandwidth while requiring minimum operating and capital expenditures.

### 8.1   SM (Smart Manufacturing)

SM is an abbreviation for Smart Manufacturing.

The industrial industry is transitioning from automated to smart technologies. Each lifecycle step generates a large quantity of data, including product design, allocation, sale, and servicing. The data is partitioned, which makes data collecting

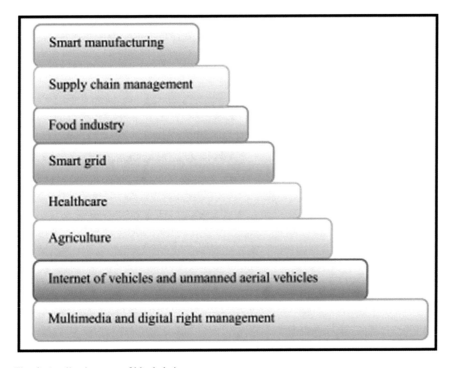

**Fig. 4** Application area of blockchain

and analytics more challenging. IoT systems are linked via a peer-to-peer network for data exchange in industrial sectors, and interoperability concerns are addressed. The use of BCoT has improved the security of smart manufacturing.

## 8.2 Management of the Supply Chain

The phrase "supply chain" refers to a group of people, suppliers, and actions throughout the product lifecycle. It begins with the creation process and progresses to the sale process, beginning with the supplier and ending with the manufacturer. As depicted in Fig. 5, the supply chain begins with the supplier and continues with the producer, trader, seller, and buyer.

Supply Chain Management refers to the process of managing data and finances as they change as a result of the supply chain process (SCM) as shown in Fig. 5. A product was made up of several pieces that were provided by various producers from various nations. It is costly to implement anti-fraud technology in each product.

Tamper-resistant and perceptible blockchain are used to ensure data provenance. The traceability ontology was established with IoT and blockchain technologies based on the Ethereum blockchain policy. The BCoT reduced service expenses. The

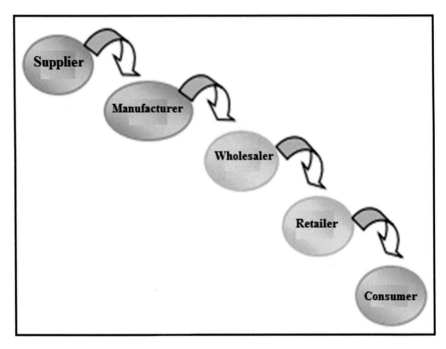

**Fig. 5** Supply chain management process

rapid integration of blockchain and IoT reduced costs and risks. The blockchain-based machine learning platform facilitates secure data exchange among many efforts to improve customer support.

(a)    Industry of food

In the food sector, BCoT improved the product life cycle. Traceability of food items is critical for ensuring food safety. It is a difficult task to provide food traceability for IoT. Several vendors were included in a food firm. For food processing, traceability necessitated raw material data from suppliers. Blockchain ensured data origin and traceability in the food business. RFID and blockchain are being used in China to build a supply chain platform from farming to food manufacturing. The technology ensured the data traceability of the food supply chain. Through the provision of traceable food items, blockchain technology improved food safety.

(b)    Smart grid

The appearance of distributed renewable energy resources changes when energy customers transition from pure shoppers to prosumers. Energy prosumers have extra energy sell it to other customers. P2P energy trading is the exchange of energy between a prosumer and a consumer. It is difficult to provide safe energy dealing in a dispersed atmosphere. Blockchain technology made it possible to provide safe peer-to-peer energy trade. Through blockchain consensus, an energy trading system based on consortium blockchains was created to minimize costs without the use of a broker. To maintain the confidentiality of transactions, a decentralized energy trading system based on blockchain was created.

(c)  Health care

Because of population, health care is one of the most pressing social-economic concerns. It featured increasing demands for healthcare services as a result of insufficient facilities. The advancement of wearable healthcare devices and the availability of healthcare data give the potential for delivering remote monitoring services at home. With heartbeat, diabetes, and blood pressure monitoring, wearable gadgets decide and collect healthcare data. Doctors may access healthcare data at any time and from any location thanks to networks. The susceptibility of healthcare equipment poses several problems in terms of preserving privacy while increasing data security.

(d)  Vehicle internet and unmanned aerial vehicles

Vehicle-to-vehicle networks, vehicle-to-roadside networks, vehicle-to-infrastructure networks, and vehicle-to-pedestrian networks are all part of the Internet of Vehicles (IoV). To solve security issues, the blockchain is linked with IoV. In IoV, blockchains were subject to a trust-management policy. Their message responsibility is allowed by PoW/PoS consensus implemented using RSUs. Blockchain technologies are used in smart grids to preserve energy and communicate data between electric automobiles and hybrid electric vehicles.

(e)  Agriculture

To increase the number of agricultural goods, smart agriculture used modern technology such as IoT, GPS, and bigdata. Information is saved in the control system and is analyzed by AI. The integration of information technology in smart agriculture reduces the cost of the agricultural supply chain. Distributed Ledger Technologies (DLTs) have improved the efficiency, traceability, and simplicity of agricultural supply chains. Two related structures must be created to minimize repetitive problems in accumulating data on the blockchain.

(f)  Digital and multimedia appropriate management

Media distribution is a type of digital multimedia material dissemination that includes audio, picture, and video. The advantages of traditional internet content distribution mediums were improved accessibility, lower costs, and greater performance. Cloud-based Content Delivery Networks (CDNs) are selected due to their low housing costs. The planned systems had inherent issues that were difficult to overcome. Blockchain-based RIGHTs management system (BRIGHT) based on blockchain video files. Existing multimedia failed to save any data about ownership or media adaptation information.

# 9   Economic Impact of AI-Enabled Blockchain in 6G Based Industry

During 2020 and early 2021, 6G research efforts were more popular as governments across the world began to look for new technologies before competitors. This may be divided into several recent major investments.

The nation has already sent a 6G experimental satellite into orbit, according to China's official news agency. It is believed that the satellite was one of 13 new satellites deployed by China during the November 2020 long-march-6 rocket launch period. The China Global Television Network reportedly stated that the satellite weighed 70 kilos and was tested on long-distance terahertz data. The satellite may be used to monitor agriculture, forest fires, and other environmental information. Recently, CNIPA has revealed that it holds 35 percent of around 38,000 patents in 6G of China's national intellectual propriety agencies.

The 6G Flagship research initiative in Europe is attempting to bring together 6G technology research, now focused at Oolu University, Finland.

In the next two years, Japan will spend 482 million dollars to enable 6G expansion. This grant will also provide an opportunity to develop wireless initiatives by researchers. The ultimate objective of the country is to show the prominent mobile technology by 2025.

In Germany, in 2021, Vodafone Germany announced the establishment of a Dresden 6G Research Centre.

It is little wonder Samsung worked with 6G in South Korea, which believes sophisticated technology like holograms to be particularly promising. They are another company that predicts the first 6G deployment as early as 2028.

In Russia, the NIIR Research & Development Institute and Skolkovo Science and Technology Institute have published a 2021 estimate that 6G networks can be made accessible as early as 2035.

For the United States, the 6G endeavor is more private than the government-sponsored effort, however, in 2021 the federal government announced a 6G research collaboration with South Korea. Some cellphone carriers are advancing their own 6G development in America.

The major industry initiative with ATIS named the Next G-Alliance is, in particular, AT&T, Verizon, and T-Mobile to organize and continue research in 6G in North America. In May 2021, a technique-based work program was started by the Next G Alliance to organize a series of new working groups to create the technology for 6G. The US is second behind China with around 18% of all 6G pa, if patent numbers are true.

## 10 Conclusions and Future Application

With the rapid growth in the number of connected IoT devices, several barriers arise to limit IoT adoption across various applications. There are several worries regarding interoperability as solutions are deployed, resulting in new data silos. The centralized design of an IoT solution necessitates that IoT device owners trust the organizations to keep their data secure. Blockchain is a promising technology that has aided in the resilience of IoT networks.

A distributed ledger avoids the difficulties of centralized design and saves data through its features. Blockchain built trust between IoT devices by reducing the danger of tampering via blockchain cryptography. It reduced costs by removing the middlemen and intermediaries' overhead. Although blockchain provides a solution to many IoT difficulties, every merger of two embedded technologies introduces additional issues and hurdles. Because of IoT needs such as security, data privacy, consensus protocol, and smart contracts, the characteristics of blockchain are changing. Several research articles on blockchain integration with IoT are examined to investigate security concerns.

The degree of data confidentiality and integrity is likewise poor in certain existing efforts. The scientific hurdles and unresolved concerns in integrating blockchain with a 6G communication network are examined. Following that, future research recommendations for blockchain-enabled IoT with a 6G connection are presented. The article's future scope is to connect blockchain with IoT 6G technology to reduce computing costs. Depending on the expectations and requirements, 5G technologies' security and privacy problems will be mitigated.

## References

1. Popescu D, Vlasceanu E, Dima M, Stoican F, Ichim F (2019) Aerial robotic team for complex monitoring in precision agriculture. In: Proceedings of IEEE 15th international conference distributive computer sensor system (DCOSS), pp 167–169
2. Yeom J, Jung J, Chang A, Ashapure A, Maeda M, Maeda A, Landivar J (2017) Cotton growth modeling using unmanned aerial vehicle vegetation indices. In: Proceedings of IEEE international symposium on geoscience and remote (IGARSS), pp 5050–5052
3. Ruangwiset (2014) The application of unmanned aerial vehicle to precision agriculture: Verification experiments of the power consumption. In: Proceedings of IEEE international conference on information science, electronics and electrical engineering. vol 2, pp 968–971
4. Pederi YA, Cheporniuk HS (2015) Unmanned aerial vehicles and new technological methods of monitoring and crop protection in precision agriculture. In: Proceedings of IEEE international conference actual problems of unmanned aerial vehicles developments (APUAVD). pp 298–301
5. Vasudevan A, Kumar DA, Bhuvaneswari N (2016) Precision farming using unmanned aerial and ground vehicles. In: Proceedings of IEEE technological innovation in ICT for agriculture and rural development (TIAR). pp 146–150
6. Vihari MM, Nelakuditi UR, Teja MP (2018) IoT based unmanned aerial vehicle system for agriculture applications. In: Proceedings of IEEE international conference on smart systems and inventive technology (ICSSIT). pp 26–28

7. Subba Rao VP, Rao GS (2019) Design and modeling of an affordable UAV based pesticide sprayer in agriculture applications. Proceedings of IEEE 5th international conference on electrical energy systems (ICEES). pp 1–4

8. Saha K et al (2018) IoT-based drone for improvement of crop quality in the agricultural field. In: Proceedings of IEEE 8th annual computing and communication workshop and conference (CCWC). pp 612–615

9. Prates PA, Mendonça R, Lourenço A, Marques F, Barata J (2019) Autonomous 3-D aerial navigation system for precision agriculture. Proceedings of IEEE 28th international symposium on industrial electronics (ISIE). pp 1144–1149

10. Saad W, Bennis M, Chen M (2020) A vision of 6G wireless systems: Applications trends technologies and open research problems. IEEE Netw 34(3):134–142

11. Giordani M, Polese M, Mezzavilla M, Rangan S, Zorzi M (2020) Toward 6G networks: Use cases and technologies. IEEE Commun Mag 58(3):55–61

12. Al-Arab M, Torres-Rua A, Ticlavilca A, Jensen A, McKee M (2013) Use of high-resolution multispectral imagery from an unmanned aerial vehicle in precision agriculture. In: Proceedings of IEEE international symposium on geoscience and remote sensing (IGARSS). pp 2852–2855

13. Holness C, Matthews T, Satchell K, Swindell EC (2016) Remote sensing archeological sites through unmanned aerial vehicle (UAV) imaging. In: Proceedings of IEEE international symposium on geoscience and remote Sensing (IGARSS). pp 6695–6698

14. Flores DA, Saito C, Paredes JA, Trujillano F (2017) Aerial photography for 3D reconstruction in the Peruvian highlands through a fixed-wing UAV system. In: Proceedings of IEEE international conference on mechatronics (ICM). pp 388–392

15. Flores DA, Saito C, Paredes JA, Trujillano F (2017) Multispectral imaging of crops in the Peruvian highlands through a fixed-wing UAV system. In: Proceedings of IEEE international conference on mechatronics (ICM). pp 399–403

16. Samad T, Bay JS, Godbole D (2007) Network-centric systems for military operations in urban terrain: The role of UAVs. Proc IEEE 95(1):92–107

17. Prajapati HP, Yatharth B, Singh M (2017) Imaging of crop canopies for the remote diagnosis and quantification of field responses. In: Proceedings of IEEE international conference on inventive systems and control (ICISC). pp 1–5

18. Tetila EC, Machado BB, de Souza Belete NA, Guimarães DA, Pistorius H (Dec. 2017) Identification of soybean foliar diseases using unmanned aerial vehicle images. IEEE geoscience remote sensing letters 14(12):2190–2194

19. Honrado J, Solpico DB, Favila C, Tongson E, Tangonan GL, Libatique NJ (2017) UAV imaging with low-cost multispectral imaging system for precision agriculture applications. In: Proceedings of IEEE global humanitarian technology Conference (GHTC). pp 1–7

20. Hassan-Esfahani L, Torres-Rua A, Ticlavilca AM, Jensen A, McKee M (2014) Topsoil moisture estimation for precision agriculture using unmanned aerial vehicle multispectral imagery. Proceedings of IEEE geoscience and remote sensing symposium pp 3263–3266

21. Chouhan S, Choudhary S, Ajay Rupani TU, Whig P (2017) Comparative study of various gates based in different technologies. Int J Robotics Automat 3(1):1–7. ISSN: 0826–8185. (Scopus)

22. Whig P, Ahmad SN (2017) Signal conditioner circuit for water quality monitoring device using current differential transconductance amplifier. Inf Technol Elect Eng J 6(2). ISSN No. 2306–708X

23. Whig P, Ahmad SN, Priyam A (20182018) Simulation & performance analysis of various R2R D/A converter using various topologies. Int J Robotics Automat 2(1):128–131. ISSN: 0826–8185. (Scopus)

24. Whig P, Ahmad SN (2018) Novel Pseudo PMOS Ultraviolet Photo Catalytic (PP-UVPCO) Sensor for Air Purification. Int J Robotics Automat. (Scopus)

25. Whig P, Ahmad SN (2019) Novel photo catalytic sensor output calibration technique. SSRG Int J VLSI & Signal Process (SSRG-IJVSP) 6(1):1–10

26. Whig P, Rupani A (2019) The development of big data science to the world. Eng Rep 2(2):1–7

27. Kulu N, Başkaya M, Keleş A, Altan A, Hacioglu R (2018) Determination of fruit health status and yield with unmanned aerial vehicle. In: Proceedings of IEEE 2nd international symposium on multidisciplinary studies and innovative technologies (ISMS). pp 1–4

28. Ekiz, Arıca S, Bozdogan AM (2019) Classification and segmentation of watermelon in images obtained by unmanned aerial vehicle. In: Proceedings of IEEE 11th international conference on electrical and electronics Engineering (ELECO). pp 619–622
29. Rappaport TS et al (2019) Wireless communications and applications above 100 GHz: Opportunities and challenges for 6G and beyond. IEEE Access 7:78729–78757
30. David K, Berndt H (Sep. 2018) 6G vision and requirements: Is there any need for beyond 5G? IEEE Veh Technol Mag 13(3):72–80
31. Singh R, Tanwar S, Sharma TP (2020) Utilization of blockchain for mitigating the distributed denial of service attacks. Security Privacy 3(3):e96
32. Gupta R, Tanwar S, Kumar N, Tyagi S (Sep. 2020) Blockchain-based security attack resilience schemes for autonomous vehicles in industry 4.0: A systematic review. Comput Elect Eng 86
33. Gupta R, Tanwar S, Tyagi S, Kumar N, Obaidat MS, Sadoun B (2019) HaBiTs: Blockchain-based telesurgery framework for healthcare 4.0. In: Proceedings of IEEE international conference on computer, information, and telecommunication systems (CITS). pp 1–5
34. Hathaliya J, Sharma P, Tanwar S, Gupta R (2019) Blockchain-based remote patient monitoring in healthcare 4.0. In: Proceedings of IEEE 9th international advanced computing conference (IACC). pp 87–91
35. Gupta R, Shukla A, Tanwar S (2020) AaYusH: A smart contract-based telesurgery system for healthcare 4.0. In: Proceedings of IEEE international conference of communication workshops (ICC Workshops). pp 1–6
36. Luu L, Chu D-H, Olickel H, Saxena P, Hobor A (2016) Making smart contracts smarter. In: Proceedings of ACM SIGSAC conference on computer and communications security. pp 254–269
37. Tsankov P, Dan A, Drachsler-Cohen D, Gervais A, Buenzli F, Vechev M (2018) Security: Practical security analysis of smart contracts. In: Proceedings of the 27th ACM SIGSAC conference on computer and communications security. pp 67–82
38. Tikhomirov S, Voskresenskaya E, Ivanitskiy I, Takhaviev R, Marchenko E, Alexandrov Y (2018) SmartCheck: static analysis of ethereum smart contracts. Proceedings 1st international workshop on emerging trends in software engineering for blockchain. pp 9–16
39. Kakavand H, Kost De Sevres N, Chilton B (2017) The blockchain revolution: an analysis of regulation and technology related to distributed ledger technologies. In: Akram SV, Malik PK, Singh R, Anita G, Tanwar S (2020) Adoption of blockchain technology in various realms: opportunities and challenges. Security Privacy, vol 3, no 5, pp e109
40. Bodkhe U et al (2020) Blockchain for industry 4.0: A comprehensive review. IEEE Access 8:79764–79800
41. Global Drone Package Delivery Market Report 2019–2030: Industry Dynamics Trends and Connected Use Cases, Apr. 2020
42. Chen S, Liang Y-C, Sun S, Kang S, Cheng W, Peng M (Apr. 2020) Vision requirements and technology trend of 6G: How to tackle the challenges of system coverage capacity user data-rate and movement speed. IEEE Wireless Commun 27(2):218–228
43. Manesh MR, Kenney J, Hu WC, Devabhaktuni VK, Kaabouch N (2019) Detection of GPS spoofing attacks on unmanned aerial systems. In: Proceedings 16th IEEE annual consumer communications & networking conference (CCNC). pp 1–6
44. Whig P, Ahmad SN (2016) Simulation and performance analysis of low power quasi floating gate PCS model. Int J Intelligent Eng Sys 9(2):8–13. (Scopus). ISSN: 2185–3118
45. Whig P, Ahmad SN (2016) Ultraviolet photo catalytic oxidation (UVPCO) sensor for air and surface sanitizers using CS amplifier. Glob J Eng Res Eng F 16(6):1–13. ISSN Numbers: Online: 2249–4596 Print: 0975–5861, doi: https://doi.org/10.17406/GJRE
46. Sinha R, Prashar S, Whig P (2015) Effect of output error on Fuzzy interface for VDRC of second-order systems. Int J Comput Appl 125(13). ISSN: 0975–8887
47. Rupani A, Deepa, Gajender, Whig P (2016) A review of technology paradigm for IoT on FPGA. Int J Innov Res Comput Commun Eng 5(9):61–64. ISSN (Online): 2320–9801/ ISSN (Print): 2320–9798

48. Whig P, Ahmad SN (2016) Simulation and performance analysis of multiple PCS sensor systems. Electronics 20(2):85–89. (Scopus). ISSN: 1450–5843
49. Whig P, Ahmad SN (2016) Modelling and simulation of economical water quality monitoring device. J Aquacul Marine Biol 4(6):1–6. (Scopus). ISSN: 2378–3184
50. Whig P, Ahmad SN (2017) Controlling the output error for photo catalytic sensor (PCS) using fuzzy logic. J Earth Sci Climate Change 8(4):1–6. (Scopus). ISSN: 2157–7617

# Comprehensive Evolution of Pharmaceutical Industries by Pioneering Blockchain Technology with 6G Wireless Networks Amalgamation

**Firdous Sadaf M. Ismail and Sadaf Gauhar M. Mushtaque**

**Abstract** When it comes to the use of new technology, the pharmaceutical industry has undergone many transformations. Operational issues and counterfeit drugs are two of the most commonly encountered challenges in the pharmaceutical industry; the solution is to provide authenticate vigilance to track pharmaceutical products from starting of production to consumption for avoiding further financial losses. Due to blockchain's streamlined nature for keeping total confidentiality and efficiency to the traditional method; it has the potential to resolve it and speed up the process by emerging 6G wireless network. Despite the fact that blockchain-based systems offer many benefits across the pharmaceutical industry, the adoption rate of blockchain technology remains somewhat low due to the lack of knowledge, awareness, and the lack of understanding that why, how, and where blockchain technology with 6G is required in their domains. This chapter explains blockchain technology, the benefits of a blockchain-based system and how it enhances the overall efficiency of pharmaceutical industries by integrating 6G wireless networks.

**Keywords** Blockchain technology · Pharmaceutical industry · Sixth Generation (6G) wireless networks · Supply chain management · Distributed ledger technology

## 1 Introduction

### 1.1 An Overview of the 6G Wireless Network

6G wireless networks are designed to provide future data-intensive smart societies with complete automation by integrating wireless network characteristics from the ground, air, space and undersea. 6G is anticipated to achieve 607 Exabytes for each month around 2025 and 5016 Exabytes around 2030, due to substantial volumes

F. S. M. Ismail (✉)
Department of Computer Science & Engineering, GNIT, Nagpur, Maharashtra, India
e-mail: firdoussadaf2810@gmail.com

S. G. M. Mushtaque
Department of Applied Science and Humanities, GNIT, Nagpur, Maharashtra, India

© The Author(s), under exclusive license to Springer Nature Singapore Pte Ltd. 2022     225
M. Dutta Borah et al. (eds.), *AI and Blockchain Technology in 6G Wireless Network*,
Blockchain Technologies, https://doi.org/10.1007/978-981-19-2868-0_11

of data traffic and new multimedia technologies. There's a need to hook up a vast group of ubiquitously connected heterogeneous devices (e.g. pharmaceutical manufacturing organizational structure, along with the internet of things, to help enable a variety of network facilities) is driving the need for 6G communication systems to be almost virtualized, software designated, and cloud-based structures [1, 2]. However, this goal is hampered by a variety of trust-related challenges that are typically disregarded in network architectures. Blockchain, a fresh and imaginative technology that has arisen in the previous decade, provides a feasible solution. Because of its decentralization, transparency, anonymity, immutability, traceability, and resiliency, blockchain can foster cooperative trust within and between separate network elements and promote, for example, effective resource sharing, trusted data interaction, secure access control, privacy protection, and tracing, certification, and supervision functionalities for wireless networks [3].

## 1.2 Blockchain Technology

Blockchain technology is an electronic ledger or series of blocks connected together via encryption. These blocks include data in the form of the block header and the block (which contains data); the coupling of blocks forms the chain, which contains knowledge regarding the current block as well as the address of some other blocks (Fig. 1). Blockchain creates a data structure through the orderly connection of nodes. This approach was invented in 1991 as the "digital documentation" of such contents. Initially, the application of blockchain technology was highly limited and quite infrequent. With the debut of bitcoin, a platform for digital money (cryptocurrency) invented by Satoshi Nakamoto, which advanced the usage of blockchain technology, blockchain technology attracted extensive attention in 2009 [4]. A blockchain technology is a form of database that is built to be read only once, according to its designs and specifications. Data is maintained in a blockchain decentralized network and is a transactional sort of information, therefore the owner owns the private key to ensure that no one else may access it [5]. In addition, the owner has access to the Inter Planetary File System (IPFS). When it comes to health care, the operation of blockchain technology is the reality that the extension of traditional healthcare information systems includes a variety of operations such as, but not limited to, having recovery mechanisms in place, performing backup storage services, and clinching up-to-date fields. Since the information in blockchain technology is disseminated over the whole linked network with no single point of failure, an entrenched backup mechanism is created [6]. Furthermore, a solo version of content is duplicated across every node of the blockchain (Fig. 1), which minimizes the volume of interactions involving information systems and thus factors for decreasing the pressure on the healthcare environment.

A blockchain can be stated as the spreading technology when accuracy and security of digital data are the big concern with respect to growing data all over the world considering today's expanding digital era [7]. Blocks in blockchain are the

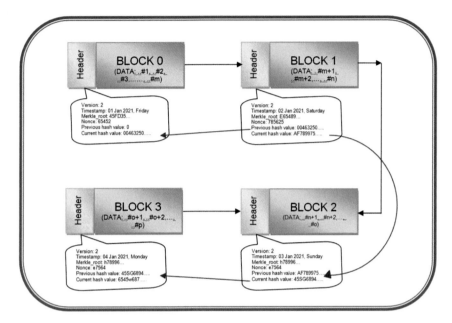

**Fig. 1** Blockchain's basic structure

growing index of data and these blocks are logically linked together with each other using cryptography. Maintaining adequate accuracy and preventing the degradation of critical data; each block contains a timestamp (to validate the existence of the transaction), a cryptographic hash (hashing value for every connected block is much like a "digital fingerprint" that guarantees block confidentiality and is always unique for each block for supplying encrypted data and reliability), and transaction data [8]. Generally, Merkle tree is used to represent transaction data. One more benefit is data modification resistivity since each block contains the knowledge about the block which one is previous to it and concertedly all form a chain as shown in Fig. 1. The block after each additional block fortifies its predecessor block, this strictly restricts the alteration of data without altering all subsequent blocks. Peer-to-peer channels are used to administer and govern blockchains, which serve as just a publicly distributed ledger [9]. Blockchain is a publicly distributed ledger in which nodes follow a mechanism to validate and interact with new nodes. There is no need for a centralized server or system to store blockchain data. The distributed data was carried out by millions of blockchain machines all around the world. Data may be notarized on these computers since it resides on each and every node and is openly able to be verified. Blockchain technology is built on three pillars: decentralization, transparency, and immutability. The brief description that how blockchain works and information flows is shown in Fig. 2.

**Fig. 2** Fundamental
operations of blockchain

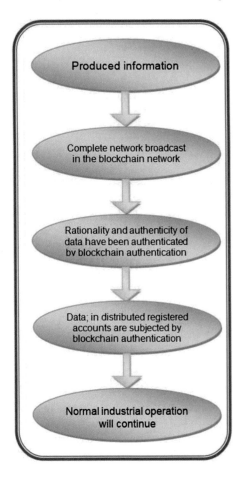

### 1.2.1   Technology Background of Blockchain

The background of blockchain technology has origins in the crypto-currency field and
it was used originally to establish bit-coin (the first practicable digital currency) as
earlier stated in the chapter's introduction part by the author. Blockchain as distributed
ledger technology (DLT) accredits the evidence of ownership with moreover the
pass on of ownership from one individual entity to another without concerning the
centralized server or systems. This allows parties to inspect and validate transactions
independently and quite affordably [10]. Blockchain databases (describes as a value-
exchange protocol) are authenticated by mass participation and driven by people. A
blockchain may consist of numerous logical levels, as mentioned below:

- From the genesis block to the present node, the main chain has the longest
  succession of blocks;
- Infrastructure, which entails the creation of hardware blockchains;

- Orphan (Unused) blocks (its existence is outside of the main chain);
- Socializing (discovery of nodes, dissemination of information, and verification);
- Agreements like proof-of-work (PoW) and proof-of-stake (PoS);
- Information about the blocks and the transactions;
- Application (if appropriate, smart contracts/decentralized apps).

The concept of blockchain technology appeared in actuality across a white paper, which was written by Nakamoto as stated before. Nakamoto's main plan was to establish a permissionless system capable of solving the double-spending challenge by leveraging peer-to-peer distributed ledger technology to produce computational proof of transactions in a sequential manner [9]. As previously said, blockchain refers to a network of blocks, each of which stores a gathering of information about the history, present, and future. As quickly as a block enters the system and becomes a member of the chain, it plays a critical function in connecting with the previous and subsequent blocks. Each block's critical role is to distribute, verify, and record transactions across other blocks [11]. This also has the advantage of not allowing a block in any chain to be changed or deleted because doing so would lead to significant change in every subsequent block. As a result, the blockchain network is a decentralized information system that contains information about all previous transactions and is managed by a preselected protocol that interprets the direction of validating and completing transactions, as well as the functioning of its members and the entire network. Furthermore, because the information is held on each node controlling each of the various networks, this sort of network is sometimes referred to as a distributed register. A transaction group in blockchain networks combines blocks of transactions interlinked on the chain by using hashing of the preceding block's record. As a result, the core security element of different blockchains, a particular set of characteristics, is imposed. Therefore, more information is provided to a block by the edge of the link (the older it is), the more it is protected against alterations. Even when an attacker tries to change any of the keys, the regional registry will become invalid immediately. This result will be obtained also because hash values are stored domestically within the next block's header so will be completely different depending on the hash function methods [12].

### 1.2.2 Categories of Blockchain Technology

Three blockchain-technology categories have indeed been investigated in audience access types based on: public, private, and consortium. Anyone can gain inclusion and participation in yet another transacting party using public blockchain technologies. Therefore, it does not imply that transactions would have limited confidentiality because public blockchain uses consensus techniques [13, 14]. Private blockchain technology allows only authorized parties to connect to a network. Rare beneficiaries can sometimes be constrained for those that have been preapproved over an acceptable blockchain. Furthermore, multiple levels of access to the data in the database can also be allowed to individuals. Private blockchain incorporates a high

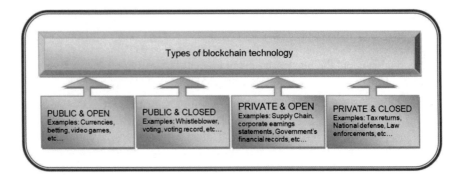

**Fig. 3** Categories of blockchain technology in users' access level

degree of security and privacy. A sole central authority determines the consensus mechanisms in the private blockchain [13, 14]. Consortium blockchains facilitate most users to access content, but then only a few have the right to write toward the blocks. Consortium blockchain encourages openness, partnership, coordination, and promotes loyalty [15]. Similarly, there are conceptions of open and closed blockchain technology considerably away from public, private, and consortium blockchain technology. The terminology public and private blockchain indicates the level of accessibility that users have to insert data to the blockchain, while open and closed blockchain refers to the amount of access that users have to read the data from the blockchain [16]. Certain legitimate uses are derived in Fig. 3 show a few uses for four different types of views.

### 1.2.3 Key Elements of Blockchain

The following are a few essential key aspects that are driving blockchain growth:

- Distributed: With each incoming transaction, the distributed ledger is revised and transmitted across the computers linked to the blockchain.
- Safe (Secure): There is no fraudulent activity on blockchain owing to permissions and cryptography.
- Transparent: Since every node or member in the blockchain does have a replica of the blockchain ledger, only those who are participants have accessibility including all transaction data. Participants may authenticate one other's credentials without any need for intermediaries.
- General Agreement (Consensus-based): For a transaction to be legitimate, all relevant network members must agree. This is accomplished by employing consensus techniques.
- Versatility (Flexibility): Smart Contracts that trigger depending on particular parameters may be programmed into the platform. The blockchain technology can evolve at the same rate as business operations.

## 2  Blockchain Technology in Pharmaceutical Industries

Since the patients are relying on the products which are manufactured by the pharmaceutical industries, therefore pharmaceutical companies play an important and major role in the healthcare industries [8]. The coherent goals of the pharmaceutical industries are to discover, develop, produce, and sell medical drugs (as shown in Fig. 4); as medications to patients, these drugs are used. Pharmaceutical Industries can track and trace drugs by using blockchain [17]. The pharmaceutical industry can be technologized by blockchain as it established three influential elements such as privacy, transparency, and traceability in pharmaceutical industries; for helping with the industry's protocols, privacy, practices, and global regulations [18]. The victory factor of administration of blockchain technology in pharmaceutical manufacturing is individually explained below [19]:

- **Trust and Transparency Factor**

  The correlation of trustworthiness characteristics that occurs between both the suppliers as well as the consumers is quite important in communication. Trust can motivate purchasers to acquire based on fair and concise information about the product, input validation correctness, responsibility, and the stage of regulatory oversight towards the transaction process, which improves security and the agility of transactions among producers and users. The connection of authenticity between suppliers of pharmacological raw materials as well as the industries may be preserved by both parties through transparency that takes place as a result of the approach, which could be pursued by both the providers and the industrial side. The transparency characteristics of blockchain technology are one of the most crucial aspects that have been governed instantaneously. Certain practices are required to ensure that there would be no data alteration or exploitation

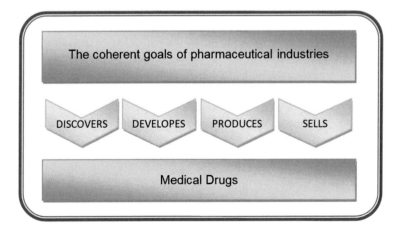

**Fig. 4**  Pharmaceutical Industries Goals

by the entities deployed in the network. Information can be shared and stored around across networks during transactions, allowing an individual transaction to somehow be traced and tracked. With many processes deliberately tried, the quality of transaction information, as well as the user's faith in the system, will increase.

- **Traceability Factors (Tracking)**

The tracking ability and traceable attributes of blockchain methods can help to have a fuller source of knowledge. An intelligent contract framework is also being recruited throughout the method of information storage and archiving, which ultimately resulted in an effortless comprehending record tracking beginning only with the original manufacturing process inside the firm and ending well with user parts [20]. Track-and-trace principles may also be used to restrict and decrease transactions in order to combat the dangers of dissemination variations and counterfeit pharmaceuticals that have been found in the pharmaceutical corporation's procurement.

- **Security and Real-Time Factor**

Because the blockchain network security will be out through all transaction processes prohibited by employing encryption algorithms, security is indeed a very sensitive matter and a concern that has always been exposed throughout the infrastructure. The network participants who do not need to be willing to take part in the network may conduct their transactions besides being affected in that too, independently of the danger, and the interested parties would embrace this system and feel more comfortable [21]. Transactions that are in real time are forced to carry out all the control processes in the pharmaceutical companies' transaction workflows. The transactions typically pop up in real time amid administrative duties allow most relevant parties to only be implicated in evaluating when and where it is, delivering transparency in the relevant data and perhaps keeping costs down. As a result, it fosters deeper goodwill among all parties to the agreement, particularly the eventual consumers.

- **Provenance Data and Immutable Factor**

The approach will allow for perhaps a validation of such transaction data that occurs inside the channel having details that could be recognized and monitored; all such data can be retrieved and thus are identified as dependable sources, transparent records with safe operation. With the proofs of transaction data, the network's collaborating members are gradually becoming enlarged to establish transactions with one another. Blockchain technology establishes a transaction process that's also transferred across peers and stopped by a verification procedure before being integrated with later transactions &encoded using cryptographic primitives. As a result, zero data items may be reformed. This unchangeable element is what allows the blockchain system to gain the confidence of the network's current users.

## 2.1   Research Background of 6G

Blockchain techniques have gained headway for evenly distributed purposes within the context of 6G connectivity, and therefore are recognized as being one of the fundamental underlying trustworthy innovations that already has attracted a huge interest from both the academia and the industry. Blockchain offers a comprehensive and decentralized framework for permanently preserving data and certifying transactions across different organizations without involving central midfielders. Blockchain incorporation alongside 6G wireless networks has the potential to help MNOs precisely monitor and report resource and spectrum use, along with providing additional rewards including enhanced spectrum auctions, cheaper general and administrative costs, and sector sharing. Blockchain may measure real-time resource consumption and boost spectrum utilization by instantaneously assigning spectrum in response to demand because of the inherent advantage of transparency [22]. Throughout the recent few years, a modest variety of blockchain research programs on 6G communications have surfaced; as more of a result, industry-oriented research is still very much in adolescence. Blockchain, in the case of smart industry sectors, performs a crucial function in facilitating member nodes to transmit data in an encrypted manner, in addition to the characteristics of authorization, traceability, and remote monitoring. Researchers estimate that even after the fatal epidemic, productivity necessity for real-time and virtual engagements will increase, as will the amplification of numerous different huge computational heavy verticals including cashless economy using crypto currencies, factory automation, virtual reality, and device-to-device connectivity (D2D) will surge. In order to meet the anticipated demand rise, a number of key obstacles in the domain of 6G communications have been identified. The distributed blockchain ledger technology is one of the most important facilitators for overcoming limits and improving operational standards [23].

## 2.2   Research Background and Use Cases of Pharmaceutical Industries

The pharmaceutical is a $482 (referred from an official source) industry and is considered as a critical part of health care. The research on the pharmaceutical industry concludes that it is not in good condition. In reality, almost 90% of the drugs break down to reach the clinical trial phase suggesting it is the intensity of breakdown. From this, the conclusion can be extracted that novel drugs rarely reach the FDA approval processes because insufficiency of patients' data; by using which the pharmaceutical companies base their research phases and approval processes. Periodically, pharmaceutical industries require dealing with the returned drugs. The major challenge is the returns are comprised of forged drugs [24]. Therefore, pharmaceutical companies urgently need to identify and then separate, before selling of such returned drugs to the market. Drugs should be barcoded and serialized to make them safe from compromised returns; decentralized blockchain utilization is the solution. For this reason, the package serial number on the blockchain can be easily recorded by pharmaceutical manufacturers, by which drugs can be verified and validated from anywhere. Blockchain pharmacy, therefore, reasons arise of a new era.

### 2.2.1   Analytics

- **Records/Data**: Better analytics require access to the correct data. Because of the critical nature of clinical information, records are more important than algorithms. Scientists and analytics firms may quickly obtain the data they need for analysis via blockchain-enabled HIE (Health Information Exchange). Additionally, if the provenance of the collected data is tracked via blockchain, trust in the quality and attribute of the data grows. Blockchain can monitor not just data but then also analytical processes, such as training and validation of prediction methods. Such surveillance function can allow healthcare providers and governing agencies like the FDA (Food and Drug Administration) to have much more faith in prediction models, improving their chances of being employed in the health context.
- **Model**: To produce a final robust model, several model predictions may be pooled via a blockchain network; aggregate forecasts are generally more accurate than individual predictions. Incentives like crypto-assets and reputational ratings may also be recommended to support collaboration and improved model submissions.
- **Computation**: The request for computer resources is rising continuously as the amount of data and modeling complexity grows. Research organizations and analytics firms typically use their own computing infrastructure or use Amazon, Google, or Microsoft's computing services. Decentralized computing has become the third alternative since the emergence of the blockchain. Nowadays, a few interesting open-source peer-to-peer computation systems have emerged, such as Golem [25], SONM [26], and others. They've produced a few early items and also have a steady, if small, customer base. Maintaining confidentiality while doing computations upon this is among the most difficult difficulties in decentralized computation, so it is especially crucial in healthcare. The blockchain-based ideas, on the other hand, are already in their initial phases and must show their promises in practice.

### 2.2.2   Use Case-I: Pharmaceutical Supply Chain

The pharmaceuticals supply chain is indeed the method whereby a medicine is delivered in an adequate amount and at a reasonable cost from the vendor to hospitals, pharmacies, and patients. The pharmaceutical industry's supply chain is complicated, with medications passing via distributors, re-packagers, and wholesalers prior to actually reaching the buyer. Manufacturers have very little transparency over the legitimacy of their products across the supply chain. The counterfeit medication issues, as well as pharmaceutical arbitrage prospects, are two major challenges inside the pharmaceutical supply chain. Healthcare items some of which are offered well with the goal to misrepresent their source are known as counterfeit medications. Counterfeit medications may have insufficient amounts of active substances, unlabeled chemicals, or no active component at all. The cost of counterfeiting medicine is substantial, and it may lead to treatment failure, medical issues, poisoning, fatality, financial loss, and even a crisis of confidence there in the healthcare system. Pharmaceutical arbitrage is an issue that occurs whenever authorities, for instance, provide funding for costlier yet life-saving treatments to somehow be supplied at a cheaper price through a plan to enlarge patient access to healthcare. When medications are obtained at a cheap cost with the purpose of selling them to patients at a massive cost, arbitrage possibilities are utilized. Pharmaceutical arbitrage reduces reasonable prices, obstructs access to medications and health care, and costs the government funds [27].

Making the blockchain a promising base for increasing faith as well as transparency, users will be able to trace pharmaceutical items along the supply chain. A barcode might be detected on a drug's package at every moment that medication's ownership switches. Only trustworthy entities are allowed to make changes to the blockchain. The record is provided on real-time basis on blockchain. Producers and end users can study the history by scanning the barcode. The infrastructure should, in theory, provide drug classification, traceability, verification, and communication in the event that an unauthorized drug is discovered. End-to-end monitoring is ensured by the clever blockchain-based management system. It will prevent the possibility of illegible medication alteration, especially at gateways and other places wherein physical interaction is required.

### 2.2.3  Use Case-II: Pharmaceutical Clinical Trials

Multiple parties are involved during the clinical trial, which adds expenses and complexity: patients, pharmaceutical firms, Contract Research Organizations (CROs), hospitals, clinicians, and regulators. The numerous parties involved may not trust one another, particularly when it comes to data collecting and data monitoring systems. Blockchain could provide a solution and serve as traceability for pharmaceutical businesses and auditors. The data is uploaded to the blockchain in real time, increasing transparency. Because the data cannot be changed, trust is established. Instead of relying on third parties to audit the data, the blockchain solution employs a consensus method that eliminates the requirement for external auditors to authenticate the transaction. Code auditors are instead required. Finally, the data owner and the legible data consumer could share data on a peer-to-peer basis. Because of the blockchain solution's capabilities, the blockchain solution intrinsically provides data integrity and generates a new governance system that all participants are required to obey [27].

A blockchain framework can function as a global ledger system; allowing patients to enter their data once on a central shared database and exchange it with different potential recipients such as CROs. Patients have complete control of the data. The data purchaser may view the accessible data via a public ledger without being able to detect a patient. If the patients want to offer a data purchaser access to additional knowledge on the private blockchain, the owner may provide it at any time. Because of the shared data record that serves as a single source of truth, the data is saved and unchangeable on the blockchain throughout the clinical study. Because of the blockchain's time-stamped data collection and data flow, defined access privileges and confirmed identities may enter the system. Any modification at any time is exactly accurate and trackable. Precision medicine and analytic tools could lead to significant advancements in clinical trials once a digital database and security are built within the network of partners using a blockchain solution. To make this happen, collaboration in the form of industry alliances with policymakers is essential. As a result, a blockchain solution might potentially reduce the length of scientific cases and save money in the long run.

### 2.2.4    Use Case-III: Pharmaceutical Waste Management

A waste management system is essential for the pharma companies, as it is the ultimate goal of sustainable supply chain processes. If junk is generated in any capacity, surveillance is extremely crucial for accomplishing outcomes relating to workplace sustainability as well as the life cycle assessment. Pharma sanitation may be significant in terms of ethical trash disposal as well [28]. There seem to be two main types of pharmaceutical waste that have been studied: pharmaceutical junk that is composed of expired or underutilized compounds, including vaccines or rather sera, and therefore is thrown away by the healthcare treatment industries and domestic family units, and pharmaceutical junk that stems from healthcare facilities, healthcare organizations, and research institutes. The trash supervision process encompasses many of the procedures that are involved in regulating and overseeing garbage from start until the end. As per the United Nations, many of these plans are classified into three major categories: The very first category of activities consists of trash collection, transportation, purification, and dumping; the second category of activities consists of managing, supervising, and governing its first category's actions; Finally, the third set of events involves reducing waste volume through all the methods of customization, repurposing, and disposal, respectively.

Blockchain technology can be used to reduce pharmaceutical waste and deal with it by implementing handling protocols. The administrator will be the first person in charge of the complete waste management system in the blockchain. Clients can create a blockchain account and request that their waste be collected from their houses through this network. Where every consumer may enroll in the blockchain simply by submitting general details such as one 's name, residence, and recognition evidence. Customers would be able to browse the blockchain once the account manager has confirmed one's licensing. Following the completion of the enrollment process, the client can use the portal to submit a request for waste collection and joins a paper detailing the sort of waste and the location where it will be collected. A variety of payment methods, as well as a cash-on-delivery option, must be available. The investigator maintains a biography mostly on decentralized peer-to-peer perpetual ledger framework and posts the findings on the blockchain [4]. The inspector also selects a trash management facility, collects client input, and takes the appropriate actions. The inspector creates a profile on the distributed peer-to-peer permanent ledger framework and sends the inspection information to the blockchain. The contractors create blockchain accounts and do the tasks assigned by the municipality. The consultant's mission is always to quantify what further junk was gathered and present the findings to the municipal government. The municipal government may indeed solicit the assistance of all other authorities, such as that of the cops, to ascertain whether or not a disaster has actually happened, start investigating the details of the incident, and make a determination to the municipal council. Subsequently, it is indeed significant to mention that the blockchain program's authorities are in charge of overseeing and legislating all of it.

### 2.2.5 Use Case-IV: Digital Identity in Pharma Industries

An ailment may have been the direct consequence related to previous chronic conditions; nevertheless, the physician's given the opportunity to make an actual assessment is governed mostly by the amount and quality of these kinds of sufferer's healthcare records that is now available. Specifically, when prior diseases occurred a number of years ago and the patient has forgotten certain facts, such as medications or medical examinations, or when the patient is unable to describe elder ailments due to a lack of medical expertise. These conditions have an impact on the present doctor's judgment, and as a result, the doctor may be unable to infer the proper information for the diagnosis. Electronic medical records have replaced physical papers linked to a patient's health data, making it easier to carry, alter, preserve, and distribute; thanks to technological advancements. The most pressing concern in this field is privacy protection, which entails safeguarding patient identities and privileges. Pharmaceutical experts may also use electronic health records to produce precise medications by having access to reliable integrated data [29].

Blockchain operates as a kind of a private framework that secures one's narratives by preserving an indestructible, decentralized, and straightforward log of healthcare records utilizing sophisticated codes. Medical records are dispersed throughout several healthcare facilities, making them difficult to access. Credit goes to blockchain technology; patients may very well have entire and secure access to information and health historical background, which must be easily cited by anyone participating in the diagnosis. To guarantee security and privacy, this access can be verified using an encryption technique. As a result, the blockchain enables speedier diagnosis and tailored treatment plans. Patients, physicians, and healthcare professionals may communicate information quickly and securely since it is decentralized [30]. Furthermore, approved pharmacies will be able to examine the drug prescribed to a patient by doctors, promoting the growth of precision medicine. As a result, the digitization of the chain connects various stakeholders while also decreasing paper-work. One such methodology can then be used to document the consequences of items on patients after they had already consumed them for data sets.

### 2.2.6  Use Case-V: Data Integrity Within Pharmaceutical Industries

The ability to provide a centralized repository among inter-linked participants is by far the most fundamental advantage of blockchain technology. It guarantees that participants agree on the nature of the data transmitted to the network, which is the primary basis for their critical agreement. This benefit provides the network with unparalleled transparency, eliminating the need for middlemen and making it perfect for sectors that require frequent verification. This just accelerates judgment by improving the flow of information and decreasing confusion. To mitigate this drawback, the pharmaceutical business requires a dependable surveillance mechanism to ensure drug details are not tampered with. This issue can be fixed by utilizing a blockchain to establish a tamper-proof mechanism that's also readable and understandable to a number of participants. Tamper-proof transaction cryptographic process pertains to blockchain's peer-to-peer networking that is using consensus mechanisms to verify the confidentiality and uniformity of each new node incorporated. Every node in distributed ledger maintains a current copy of the ledger, infused with previous transactions as well as possession details. The blockchain's block structure ensures that transactions are tamper-proof. Because the blocks are connected, any unintended modification in one causes the network to become incongruent. As a result, it refuses to accept any modifications in block transactions induced by an attack. Further to that, pharmaceutical firms adopt medicinal products with detectable specifics including the title, position, timestamp, additives, drug consumption, and adverse reactions, along with licensing agreements, in order to preserve interconnection through each phase and protect medicine with missing codes from reaching the system. Many pharmaceutical businesses employ holographic technology to ensure the authenticity of their products and combat the problem of counterfeit drugs. However, this form of packaging is costly, and holograms may be copied.

### 2.2.7  Use Case-VI: Pharmaceutical Transparency

The pharmaceutical cold chain can benefit from blockchain technology since it provides transparency and addresses the requirements of the supplier, producer, logistics, distributor, and consumer [31]. Blockchain may be used to verify that all pharmaceuticals comply with patient safety standards, and smart contracts could help with this. Agreements, nodes, state databases, and rules have been put together to form the blockchain's components. The state database is being utilized to show-case and save the present state of such data base at a specified moment. Each ledger entry depicts the current condition of the medication data. The name, amount, and expiration date of the medications are all recorded along with their price. The transactions relating to medicines make up the blocks. Verification regulation would add the medicine transaction to a ledger. Node agreements are used to create policies. Agreements, nodes, state databases, and rules have been put together to form the blockchain's components. The state database is being utilized to showcase and save the present state of such data base at a specified moment.

## 2.3 Motivation for Role of Blockchain in Pharmaceutical Industries and Need to Include 6G Wireless Networks

The impression of blockchain implementation in pharmaceutical industries can be assessed owning to the fact that it will assist the healthcare industry to save up to per year $100–$150 billion and by 2025 this accounting must bring effect. The assets can be spotted in the personnel cost, operational cost, counterfeit and support functions costs, etc. The total merging of blockchain and pharmaceutical industries can progress a long way. Fifth-generation networks cannot satisfy the exceptionally elevated IoT expectations (IoTs for pharmaceutical industries), the ultimate responsibility seems to be on the 6G network to determine the highest throughput that describes a massive increase in the establishment of innovative services and functionalities including multiple-gigabit communication rate, enhanced provenance, as well as reduced latency. Owing to the paucity of spectrum resources, efficient material administration, intelligent strategic planning, and fair implementation are required for accomplishing all objectives. According to its own intelligent adaptability and integrated capabilities, blockchain is the resolution to all of these perceptions and has recently achieved tremendous relevance and is highly vital for sixth-generation wireless networks. A tremendous opportunity for blockchain related to resource supervisors well with the support of technologies such as IoTs, D2D, and domain blockchain setups. Blockchain technology handles the formidable issues of sixth-generation wireless networks, as well as its potential benefits. As no central authority authorization is expected for settling, the use of a blockchain simplifies the system simpler and quicker. Members in a blockchain network can immediately share resources. As the result, intermediaries are minimized and a copy of the shared ledger is allotted to each participant which then minimized the transaction efforts [32]. Since blockchain data have been shared by millions of participants, which is the result that no one can temper it. By all such security authentication and confidentiality, the entire system is shielded in case of cybercrimes and frauds.

## 3   Blockchain-Based Supply Chain in Pharmaceutical Industries

When blockchain technology is integrated into the pharmaceutical supply chain then healthcare may get much of benefit. One of the biggest concerns for pharmaceutical industries is the proper management of the supply chain. The difficult task with regards to the supply chain is not only to manage it effectively but also to observe and obey the standard. When dealing with drugs, multiple parameters must be tracked [30]. These parameters involve capturing of information such as humidity, air quality, temperature range, and includes many more. If any of the listed parameters are not maintained, the drug can lose its benefit and might not be convenient for the clients. The vaccine comes under one of the best illustrations. Throughout the journey of the supply chain, vaccines need an attentively operated environment. This issue can be solved by blockchain that needs to integrate the supply chain with the IoT [33]. For such integration, a supply chain can be equipped with those devices that can track the temperature, humidity, and other necessary factors. Blockchain is fairly useful for adding compliance and governance with the supply chain because of blockchains' key features such as distributed nature, immutability, and transparency.

### 3.1   Blockchain Empowered 6G Wireless Network Architecture

Blockchain is acknowledged as something of a technological innovation that would be used to unleash the great potential of 6G by enabling decentralized network governance systems; it can be used for resource exchange and cooperation, spectrum regulation, information storage, and some other purposes too. A distributed database makes it easier to record and appraise resource exchange activity. A legitimate multi-carrier market driven on a decentralized blockchain network will increase spectrum transaction efficiency while alleviating spectrum scarcity by allowing for the administration of an underused roughly comparable frequency range. In the context of data transmission and complicated calculation, blockchain is thought to be a resource-intensive approach. Minimal research on blockchain-enabled spectrum control has been conducted, but they have all ignored the effectiveness of the block chain itself, from outline to the minimizing problem. The government's spectrum controller creates high-level restrictions and limits on blockchain-enabled smart contracts that are nearly impossible to breach [34]. The legislative organization guarantees that the spectrum is governed equally throughout the nation by harmonizing various wireless connectivity protocols, supervising transceiver strength, congestion, and supervising spectrum coordination so on.

## 3.2  Case Analysis

By using the pharma business as just an instance, suppliers make available raw materials to the pharmaceutical business, which then makes and sells pharmaceuticals to medicines firms, who then sell the stuff to healthcare institutions, and lastly, the healthcare facilities sell the medical stuff to patients. Pharmaceutical manufacturing serves as the hub of the whole supply chain (A pharmaceutical group business can be formed by combining several pharmaceutical companies). The quality of the raw materials used by suppliers has a direct impact on the quality of the medications, thus pharmaceutical firms and hospitals must be involved in the selection of suppliers. In the case of a fixed business partnership between a medicine company and a hospital. Based on pre-turnover statistical data, the medical center would then suggest a pharmaceutical distribution summary screen, as well as raw material sellers and pharmaceutical industrial plants, will actively engage within every segment's project proposal in order to maintain a sustainable and sociable business arrangement with medicine corporate. Exporters, healthcare institutions, pharmaceutical corporations, and hospitals may indeed make different choice mechanisms for each section of the distribution chain via making alliances in the supply (network) chain on the blockchain. Patients may indeed join the relevant community to learn about the full medicine development or logistics processes (the blockchain utility architecture can be changed depending on the individual qualities of the supply chain company) [35].

### 3.2.1  Data Storage Mechanism

Since the succeeding node throughout the blockchain comprises each one of the preceding block's corporate data with hashing values, information legitimacy and non-tampering are acquired. Blockchain-based supply chain framework with each company's records creating a block with the same kind of legal entities. The segment storage structure of the blockchain is depicted in Fig. 1 which shows how the data storage structure of the blockchain differs from that of a unified supply chain for manufacturing enterprises. Supply chain networks of multiple industrial businesses are logically parallel, throughout the supplier relationship chain, but instead of sequential. In conventional information technology systems, the distribution channels of diversified firms are concurrent, and therefore have no connectivity with each other. The wide range of business records of supply-chain issues are linked together using cryptographic hash references to form it all into the blockchain-based system, while also digital communication. Multiple manufacturing firms' supply network topologies may fluctuate, as well as the supply-chain topic layout is significantly more complicated [36]. Whenever blockchain-based data bandwidth rises, each economic topic will also have to hold block data within its own cloud infrastructure or lease space on some other cloud infrastructure.

### 3.2.2 Data Access Mechanism

Blockchain's incontrovertible evidence method effectively eliminated the supply-chain deception crisis caused by informational imbalance. Ignoring the fact that the blockchain might have saved most of the supply chain's corporate information, the premise for rectifying the deficiencies of information sharing is getting the associated corporate data of all other corporate agendas. Relying upon it, the study contends that distribution network verification disciplines can monopolize blockchain database access competence. For regular users, gaining access only to authorized scope's material via the blockchain sector is vital, and that might be content accessibility for a segment of a node or statistics accessibility over a certain quotation. As cryptographic hash references interconnect every block and each block's items throughout the whole blockchain, blockchain access users can easily navigate the total blockchain underneath the jurisdiction of the said blockchain network. As a possible consequence, entire data with each block of said blockchain can also be built, with block relevant information framed by different business sectors and compounds, however, with the block's continuous supply, the successive block would then copy the prior block's evidence data and make new element details to content, having allowed permission people to generate information much faster. Access-users can acquire the relevant information quickly and properly via virtual connections networks; nonetheless, the prolonged chronology of the virtual connections is in clear contradiction with the time-series data established by the blockchain [37].

## 3.3 Supply Chain Management

Multiple stakeholders with numerous important requirements are required due to the increasing complexity of pharmaceutical supply. Presently, the supply chain needs to permit multiple parties to share and update data. It should validate that such details may have to communicate with national and international regulatory bodies, transport systems, and verification systems. This put forward several challenges [38]. Clearly, many of the stakeholders are allowed to input and change the data; this adds complexity to the quality of the information stored in the databases. It becomes very difficult to validate and monitor the right information and secure against human-generated errors and missing of documentation part. Blockchain is expected to transform the pharmaceutical industry supply chain by mandating a mono-use platform for all stakeholders, offering much confidentiality, accountability, and surveillance of the end-to-end transport of services. It is expected that blockchain could transform the pharmaceutical supply chain by authorizing a single-use system for all of its stakeholders, which would provide much security, transparency, and surveillance of the end-to-end delivery of goods. Supply Chain Management (SCM) process entails a number of actions in order to convey products from their place of origin to their final destination. The Supply Chain (SC) operations are designed, planned, executed, controlled, and monitored as part of this process. The many stakeholders in the supply chain operate in divisions [39].

### 3.3.1   Integrating Sixth-Generation Wireless Networks and Blockchain Technology into Supply Chain System (SCS)

The use of blockchain in the pharmaceutical sector to secure and optimize supply chains might be extensively adopted. The main focus of any solution that may be given should be on improved compliances and benefits for both the producers and the customers. Product identification, tracking, and verification are three important needs for blockchain applications. For this reason and at the same time to grow the company, medical businesses might use RFID: Radio-Frequency Identification and IoT: the Internet of Things algorithms. The benefits of blockchain in the pharmaceutical industry supply chain business, as well as its evolving impacts on inventory management by suppliers, physicians, and others, are critical for the future growth of medicine [40] and agriculture. Individuals may use blockchain technology to get a large number of constituents to agree on a frequent, authentic set of facts in the supply chain. Financial adjustments, for example, can be made by financial institutions by selecting and documenting all trades that occur inside the system. Without a blockchain, businesses must employ a large number of people to examine their requests in order to obtain bulk discounts. However, blockchain does this task without the need for extra personnel or time, obviating the need for a new value check procedure and providing significant benefits to supply chain management.

### 3.3.2   6G, Blockchain and Supply Chain Challenges (SCC)

Although the platform is revolutionary in and of itself, it has several limitations and issues when used in application areas. The study gives a precise and in-depth examination of the non-technical and technical barriers to wireless network blockchain adoption in the distribution chain.

The primary computer, which will perform in a 6G transmission environment, will necessitate the simultaneous interaction of a broad range of devices. The network framework, including such a core network, would manage a large number of businesses in real time. The issue of accuracy of inputted data is at the intersection of technological and non-technical challenges. Data placed into a blockchain must be accurate; because blockchain technology is immutable and transparent, the user cannot simply alter or modify the record. If a supply chain partner's information is recorded in an unreliable system, the inclusion of blockchain technology may be more harmful to the user than to be beneficial. The immutability of the blockchain does not guarantee the data's quality [41].

Non-Technical Challenges

Even though RFID chips and scanners are now widely available in terms of reach and accessibility, the availability of technology does not guarantee its acceptance;

many warehouses still operate using paper at critical locations. The lack of knowledge of blockchain among corporate executives, the belief that it is a fad, and the desire to wait for wider adoption before committing to the technology are all factors working against its adoption. Even corporate executives who recognize the potential of blockchain are hesitant to invest money and time in it due to a lack of industry-wide standards and procedures. To be impactful, everyone in an effective supply chain sector must be assured of blockchain's benefits; key stakeholders must be on board and reap the importance of using blockchain. As a result, acquiring market acceptance is a major challenge.

Technical Challenges

When compared to conventional database informational content, retrieving and committing records on the blockchain takes significantly longer. It also necessarily requires a considerable rise in computational resources, with scalability posing a significant challenge. Interoperability is also required among all systems that interact with the Blockchain [42]. The required payment must be brief and flexible enough to allow cash out in any currency, it includes FIAT money too.

- **Scalability**

  When the size of the input is raised to satisfy user demand then scalability is needed. Scalability refers to a system's ability to adapt and operate. There are four different sorts of scalability solutions; on-chain scalability, off-chain scalability, consensus mechanism-based scalability and distributed acyclic graphs based scalability.

- **Interoperability**

  Regardless of the fact that blockchain implementation is rising, the detachment of blockchains in their own "distribution centers" due to a lack of interoperability standards is impeding wider adoption. Solving cooperation and cross-chain interaction issues on the general populace, partnerships, and corporate blockchains could open the way for a hyper-connected world. When it comes to general agreement concepts, transactions, and contract functionality, blockchain systems must speak the same language, which means they must have identical capabilities and feature sets.

## 4  Claims and Billing Management

In general, electronic platforms are being used to handle and charge prescription medication assertions, discount bargaining; rebates and formularies, and processing and paying prescription drug claims. All these are managed by pharmacy benefit managers that include mediator among pharmacies, drug manufacturers, and payers.

Blockchain technology integrating sixth-generation wireless networks could possibly give a lot of efficient, fast, accurate, and transparent solutions that may lower the waste, alleviate pricing variations and offer a greater customer experience. Blockchain technology along with 6G might potentially intensify the insurance authorization activity on the front end and speed up claims processing on the back end. A system based on Distributed Ledger Technology (DLT) is already being used by Change Healthcare. This system track claims throughout the life cycle for a large capacity of patients (up to 50 million transactions per day). Claim verification may be done through a peer-to-peer network using smart contracts on the blockchain, and claim processing can be done automatically following claim verification. There is no need for a skewed third-party authority. Claims fraud may be prevented in a secure blockchain system, which might speed up the claims process [43].

## 5  Quality Management

With the integration of blockchain in the pharmaceutical industries clinical trial data reliability and quality will also get enhanced. The decentralized blockchain's behavior allows the clinical laboratories to use an immutable and transparent data source where hampering the data turns to be impossible. Data verification can be done whenever required by the stakeholders since the public blockchain is used to store clinical trial data. Clinical trial data is tamper proof. The clinical laboratories can trust the information fully to carry out the outcomes of their trials. By using this system, the pharmaceutical industry can also convince more patients and help patients to participate in the trials. Because blockchain technology is only as decent as its consumers, if moderate or erroneous information is added to the chain, the only thing that can be guaranteed is that reduced, inaccurate, and minimal information will exist on the chain due to immutability and decentralization. While blockchain offers many new possibilities, it is important to assess the whole execution, including what happens to data before and after it is placed on a blockchain [44]. Interoperability solutions will need to be conscientious about the data; hold, including methods for resolving inconsistencies and giving trust to various types of data. Modern technology promises to be beneficial. The benefit of blockchain is that information is secured from unwanted alteration. This has a significant impact on the economic system. The degree of trust grows with sixth-generation wireless technology, which is expected to introduce new features such as reduced latency, ultra-high confidentiality, and maximum efficiency inter-gadget coordination. As an outcome, it inspired the reader to investigate the key issues and challenges of 6G wireless technology. Sixth-generation wireless technology will change the clarity and approach to modern conduct, user-friendly, community, business, and transmission. It is expected that sixth-generation network technology will greatly assist in the transformation of various fields. Participants in contractual relationships are prohibited from engaging in opportunistic conduct. Economic efficiency improves as a result. In the instance of the pharmaceutical business, these beneficial impacts are examined. The use of blockchain technology in pharmaceuticals allows for the tracking of medications at all stages of manufacture and the assurance of their quality. With the aid of sophisticated digital devices, blockchain technology empowers to verify the validity of recipient users and medications. As a result, the quantity of counterfeit medications on the market has decreased, and the quality of medical treatment for the general public has improved. The following are a few quality-maintenance points:

- All document circulation in pharmaceutical companies may be converted into a blockchain. This approach will improve the speed with which data content is processed, assure document circulation transparency, and prevent the risk of document loss, damage, or falsification. These benefits are established by the technology itself; the generated block cannot be modified or removed.
- Blocking ensures that data cannot be tampered with. The selling of counterfeit or inferior medications would be prohibited. This is accomplished by maintaining control over all aspects of the medication production, logistics, and distribution chains.
- A unified computerized recipe database will be established. Falsified drug trafficking will be successfully combated as a result of this. Violations such as the illicit resale of costly pharmaceuticals acquired with budgetary resources on the commercial trading market would be prohibited. Each patient will be able to verify the authenticity of medical assistance in form of treatment and guidelines as well as the number of medications the patient has received.
- Prescription medicines will not be taken into account on paper, resulting in cost savings. This login account will take the place of the online digital register entries, which can give the virtual entries too. This will considerably reduce the complexity and speed the regulation of prescription medicine release while also lowering the labor cost and decreases material expenses.
- A federal patient healthcare data registry will be established. This will make life simple, faster, & economical to innovate new medications. Divide the budgets of health insurance funds and the state government budget (in terms of money allotted for health); address other healthcare-related issues.
- Ensure that funding goes to the healthcare and medical sector, particularly in the citizenry or common people sector is transparent.
- Spending on drug quality control in the state budget should be reduced. Because of the usage of blockchain, these expenses will be reduced in absolute terms. The majority will be compensated by industry businesses interested in marketing their high-quality products as well as genuine competition (removing forgery pharmaceuticals from the market).

## 5.1　Technical Aspects of Blockchain with 6G Wireless Network to Enhance Pharmaceutical Industries Operational Work

6G networks will feature a huge network of devices that connect homes, communities, and industries with multiple data transactions. To ensure safe data transport, devices must have mutual trust. To establish such trust and security, the blockchain lets each node keep a sequential ledger (chain of blocks) in which each user who modifies data within a block is accessible to and validated by everyone else to avoid fabrication. To preserve transparency, each data block must be confirmed in relation to the prior blocks by the other participating users. The dependability component is added among all users in this manner; nevertheless, privacy is compromised because all changes are accessible to everybody. The decentralized nature of blockchain allows for greater flexibility while lowering administrative costs. In a densely linked mesh network, the blockchain provides various benefits, including security, dependability, trust, and scalability. A video conference call or mixed reality-based video streaming, for example, might be an example of a blockchain in 6G networks, where the network, in addition to providing connection, also needs all parties to verify themselves and the data being transferred using specialized blockchains. This will include data properties knowledge to analyze anomalous activities in real time. However, aside from the large device connections, there are additional problems in 6G networks for blockchain implementation, such as resource constraints on devices that limit the breadth of usage of cryptographic security methods on the device. Similarly, data packets must be subjected to a thorough examination to assess danger, which becomes a monumental effort when the number of devices is large. Security assaults are more accessible due to the lack of third-party verification, as half of the total participants are compromised [45].

Blockchain will give security to mobile edge computing nodes in the future for 6G when several devices want to keep their data in the edge device. In the case of device-to-device communication, distributed security is also required for cooperative data caching among users. Furthermore, when considering network virtualization, it is clear that supplying more slices will increase network capacity. In this regard, blockchain would give the authenticity of data in each slice as well as immutability characteristics to efficiently administer the virtual network when implemented.

## 6　Applications

Blockchain is analogous to a definite, actually involved, and a successive portion of data that is monitored by many candidates who can avoid distortion by using a 6G channel. Most candidates require confirmation and acceptance of a new section of content before adding it to the blockchain. Participants demand an observer of preceding data during the verification period. As a consequence, blockchain is widely accepted as an advanced trusted mechanism for validating 6G networks.

Applications of blockchain with 6G wireless networks in healthcare distributed ledger technology (DLT) have been reshaping the healthcare sector all around due to its immutable manner of data storage and real-time information updating capability [46]. A drug supply chain is indeed one of the significant fields where applications of blockchain are pursued as the best product of innovation. In pharmaceutical industries, there are many other potential applications for blockchain technology including facilitating (to patients, to payers, to physicians, and to pharmaceutical companies) access to medical records and prescription sharing, credentialing provider, improvement of the supply chain, quality-of-care tracking, reporting and tracking of clinical trial data, adverse event tracking and evaluation and drug pricing strategy tracking [47]. In Pharmaceutical Industries, the following are several use cases for blockchain applications.

- In the pharmaceutical industry, electronic medical records are the most common use case. This is because DLT allows for the creation of a single, longitudinal, and alter-proof patient record. All data relating to vaccinations, lab test results, treatments received, and prescription history will be kept on a ledger that is held on a decentralized peer-to-peer network, and these records will be extremely useful to the pharmaceutical industry's growth.
- Taking medications as prescribed every year, tens of millions of dollars are spent on medical prescriptions owing to poor patient compliance. Individual's medications are expensive. Through application program interfaces (APIs), which will encourage the medical prescription-taking procedure, incentives can be offered for any improvement in medicine. Information saved on the blockchain will be available to both physicians and patients, giving pharmaceutical firms a huge boost.
- Blockchain technology and 6G, which allows for easy-to-share data to be available over time in a distributed ledger, can provide tailored care and feedback from healthcare centres and patients. Pharmaceutical manufacturers and medical practitioners can import the medical histories of family members with previous prescriptions to gain a comprehensive knowledge of an individual's medical state and determine whether certain improvements are required to improve the pharmaceutical industry.
- For tracking the supply chain for medical products of pharmaceutical companies that deliver medical care equipment or medications to patients, we have to navigate via a complicated distribution network. When a drug is recalled, tracing the medication back to its source becomes challenging. When blockchain applications are used to record transactions involving medications, medical equipment, and services, a tumultuous audit trail is kept. The transaction history can be utilized to reduce the counterfeiting of drugs that have dangerous adverse effects.

# 7    Challenges and Limitations

Adopting blockchain and 6G in pharmaceutical industries has some significant challenges and limitations because it is hard to construct a credible and trustable sixth-generation wireless channel due to multiple ordered ranging from telecommunication, politics, rules and conduct. It will not suffice in the 6G channel because personal integrity will be entirely dependent on IT in addition to data transmission. The parts of validity, confidentiality, and protective measures are all inter-connected in some way or another, but at a varying scale of the next generation of the network. Every pharmaceutical supply chain stakeholder; stock core business informational content in the distributed ledger of blockchain that allows everyone to access sensitive private data leads to losing competitive advantage when many competitors exist in the same supply chain. This distributed ledger technology is one of the most radical innovation enablers for tackling most current limitations and aiding 6G functional standards. Solutions and platforms based on blockchain lack full interoperability due to the absence of standardized solutions to turn implementation, integration, and adaptability easier. The role of regulatory agencies, in blockchain-based solutions, becomes more complex and pertinent. This becomes difficult for these groups of institutions to interpret the environment and the legal limits of blockchain technology. Due to the generation of undeveloped practices and controls, recent blockchain frameworks are exposing intrinsic bugs and vulnerabilities [48].

Although certain parts of blockchain technology safeguard against unwanted access, a massive leak of private data caused by a technological flaw might result in dread of what should have been a source of hope. One intriguing issue is that data on the blockchain can only be accessed via a "key," which is a one-of-a-kind pattern of letters and integers. When a key is lost, the content containing the information accesses is irreversibly lost, and cannot retrieve back. It is unconscionable to restrict access to a lifetime's worth of medical records due to the loss of one of these keys and methods must be created to rejoin individuals with the information. Solutions reportedly in use offer backdoors to accessing blockchain-based personal information essentially repurposing the problem. A further issue is that if the federalism of a blockchain is broken, such as if one company gains majority control of servers (more than two-thirds with current enterprise mechanism), one agent can become the primary agency of agreement and reshape the blockchain, violating the immutability property [49]. To prevent this from happening, new consensus technology and government control of blockchain monopolization may be required. The following are some of the most significant logistical challenges:

- **Transparency**: This is required for overall SCM optimization and efficient resource utilization throughout the supply chain. Transparency is inextricably linked to trust, which is critical in the pharmaceutical industry.
- **Traceability**: This is required to trace product mobility across the supply chain. In the pharmaceutical industry, governments and regulatory organizations are progressively legislating for the use of traceability systems to address associated issues caused by a lack of traceability [50].

- **Accountability and Liability**: This is described as providing the customer with required information about the third-party logistics provider's storage, shipping, insurance, safety checks, social norms, close monitoring, value-added (economic benefits) operations, stock management, order management, loading and other services. Buyers demand high-quality service for the price buyers pay, thus logistics accountability is equivalent to answerability. Accountability in logistics will also have a beneficial impact on confidence. The issues raised above can be solved by combining a strong set of blockchain-based smart contracts, a logistics planner, and asset condition monitoring.

## 8 Future Research Opportunities

Incorporating a blockchain into a sixth-generation wireless channel would allow the organization through information channels to monitor and retain bandwidth utilization while also distributing resources effectively. Blockchain technology enables transparency and shared information transformation, which would benefit a various variety of industries and organizations. Recent pharmaceutical industries are patient-centric, and therefore blockchain can be enhanced for the patient-centric approach. There is a need to research in the construction of regulatory architecture and agreement for stakeholders since the visibility of blockchain data in pharmaceutical industries degrades the business standards [40]. One more very important research attribute will be security and complete transparency in the supply chain to save pharmaceutical industries from denial of services and claims of the unauthenticated source. Technology vulnerabilities, phishing scams, malware, and implementation exploit, due to lack of procedures and protocols, in moving further, show serious challenges that need to be addressed [41]. There are certain elements of implementation that have yet to be investigated. Reduce the amount of computing power required to run blockchain code on low-cost IoT edge devices and improve fault tolerance in device-to-network communication. Another area of investigation will include lightweight network protocols for device-to-device communication and network propagation, as well as techniques for achieving network consensus. As stated in many research studies, the vision statement for 6G networks stated, 6G networks would be a milestone in changing the smart cellular network inside an intelligent channel which will enhance the pharmaceutical operational work. These studies predicted that 6G networks will develop by 2030, and they highlighted a number of intriguing technologies, applications, and use cases that will be benefitted and enhanced by 6G networks [51].

# 9 Conclusion

To summarize the findings, the author concludes that blockchain and 6G incorporation sets the circumstances for higher effectiveness in the pharmaceutical industry sector. Chain of custody, track ability, openness, trustworthiness, real time, provenance data, data protection, proprietary information, cost benefits, serialization, verification, audit ability, recognition, effectiveness, spontaneously, immutable, consensus-driven, as well as sustainability are success factors in the pharmaceutical firm's deployment of blockchain-based technology. With a decentralized, transparent, regulated, and safe method, blockchain operates on the basic phenomena of no repetition and symmetry of the item. In the pharmaceutical supply chain, the blockchain idea may be used to identify fraudulent medications by maintaining required drugs' supply and demand that allow Pharma businesses for keeping track of counterfeit and unlicensed drugs. Because the drug companies are typically chaotic and splintered, the presence of numerous suppliers in the supply chain adds to drug quality control issues. The following are the inferred benefits of blockchain innovation for fostering positive network administration: reducing overall mischaracterization and discrepancies, unnecessary delays from hard copies with the integration of 6G, enhancing stock control, recognizing troubles more speedily, curbing dispatch expenditure, and ramping up consumer and trust companions. Blockchain methods are currently being used in the pharma firm in the form of beginning and research initiatives. Yet, the author predicts the emergence of localized, nationwide, and maybe global communications systems are expected future as expertise and expertise get more inexpensive, hence increasing their presence. All such solutions will be linked to the networking via distributed databases of accurate statistics from medication designers and manufacturers, suppliers, pharmacies, clinics, and consumers, as well as state and international legislative and governing bodies for the pharmaceutical industry sector. In terms of expense remedy, non-repudiation, converging among pervasive computing, and use of trace ability to check unlawful actions damaging the sustainability charter; blockchain technology is an emergent distributed guarantee. A smart contract is a fundamental component of blockchain technology that reduces negative external elements during production, improves the quality of life, promotes employability, and improves social and environmental performance across a wide variety of applications related to the pharmaceutical industry, in addition to supporting economic performance. The usage of blockchain in 6G use case applications is predicted to develop, bringing benefits to intelligent living and firm segments by reducing safety concerns and ensuring the integrity of information.

**Specification of Terminologies**

1. **Cold Chain:**
   It is a strategy, vaccine supply chain, and system of regulations that ensure the proper supply and disposal of vaccines to healthcare providers at all stages, from national to regional. The cold chain is interconnected to refrigeration equipment to provide temperature-controlled environments to store, which

allows vaccines to be preserved at recommended temperatures in order to sustain potency within the time limit.

2. **Electronic-Ledger:**
   Electronic Cash Ledger is a computer system that stores and tracks transactions that contain deposits of a taxpayer. The cash ledger isolates the information head-wise like a CGST, IGST, UTGST/SGST, and CESS. Each of these preeminent heads (SGST, IGST, etc.) are further divided into five minor heads—Tax, Interest, Fees, Penalty, and Others.

3. **Cryptographic Hash:**
   A cryptographic hash function is an algorithm that takes an irrational amount of data credential, input, and character and then produces a fixed-size output of encrypted text called a hash value or just a hash only. That enciphered data can then be stored instead of the password itself and later used to confirm the user.

4. **Read Only Once:**
   This gives the demonstration that how to set up attributes for the class as read only or write once entities and stopping them from being modified by the end-user. This shows that the behavior changes when the end-user attempts to set values.

5. **Interplanetary File System (IPFS):**
   The Inter-planetary File System (IPFS) is a protocol that defines an end-to-end information channel for saving and distributed contents in an allotted file system and give interoperable smart contracts which are used to access the data and transfer it from one computer to another one. The data transfer is faster, more secure, and economical in comparison to a server database through IPFS.

6. **Merkle Tree:**
   Merkle tree is a basic aspect of blockchain technology. It is a mathematical information frame composed of hashes of individual blocks of content which provide a summary of all transactions in that given content or block. Merkle tree is a tree in which each and every leaf node is tagged with the cryptographic hash to a bit array of a firm size of an information block and every non-leaf node is tagged with the cryptographic hash of the labels of its child nodes. Merkle tree helps to prove the consistency and content of the data. Ethereum and bitcoin both use the Merkle trees structure.

7. **Peer-To-Peer Networks:**
   P2P is for a peer-to-peer network and it is a communication over a decentralized network paradigm that contains a set of devices which is referred to as nodes that jointly cache and transfer data, with each node acting as a single peer. End-to-end (P2P) information exchange takes place in this network without the use of a pivotal supervisor or server, which indicates that all nodes have the same power and do the exact identical functions. Because of blockchain's P2P architecture, all cryptocurrencies can be exchanged globally without the use of a central server or intermediaries. A bitcoin cell can be set up by everyone who wants to participate in the system of verifying and validating frames or blocks.

8.  **Distributed Ledger:**
    Distributed ledgers are databases that are shared throughout the channel and spread across multiple geographical sites. A ledger is an assembly of monetary accounts that are spread out and handled internationally in this situation. As a result, several people in various locations and institutions hold and restructure dispersed ledgers. Participants at each network node can review distributed ledgers, and can acquire an exact duplicate of the recordings shared across the channel. The modifications are reproduced and distributed to the participants if the ledger is modified or enclosed. The database is synchronized in order to ensure that it is accurate.

9.  **Distributed Ledger Technology (DLT):**
    DLT (distributed ledger technology) has become a catch-all word for multi-party systems that work in the absence of a central operator or authority, despite the presence of untrustworthy participants. Blockchain technology is a subset of the larger DLT universe that employs specific hash-linked blocks of information which consist of a whole data structure.

10. **Cryptocurrency:**
    Cryptocurrencies are digital financial strengths in which cryptography decentralized technology ensures ownership and its transfer. Cryptocurrencies are a subset of a larger class of financial assets known as "cryptoassets," which use end-to-end digital value transfers without the use of third-party organizations for transaction validation.

11. **Bitcoin:**
    Bitcoin is intended to function as a peer-to-peer electronic cash system. It must be stable or backed by the government to function as a currency. Bitcoin is a cryptocurrency based on the P2P network concept. A pseudonymous software engineer or hacker dubbed "Satoshi Nakamoto" is credited with this invention. The main goal was to develop a transaction system that was free of a central bank or monetary authority intervention, and was based on a mathematical formula rather than "third-party trust." Payments may be made in a secure, verifiable, and indisputable manner via electronic means. The implementation of this concept necessitates the creation of a payment system that is applicable to all transactions.

12. **Value-Exchange Protocol:**
    This technology allows for the exchange and recording of digital value in a global open system that ensures that once a transaction is completed, it cannot be reversed. That is crucial. To put it another way, the blockchain protocol can be used as a proof of record for any type of digital asset exchange.

13. **Orphan Block:**
    An orphan block is one that has been solved inside the blockchain network but has yet to be acknowledged owing to network latency. A block can be solved by two miners at the same time. The block reward is given to the miner who has the most thorough proof-of-work sheet.

14. **Double Spending Problem:**

The danger of a digital currency being consumed twice is known as double-spending. Because digital content may be quickly replicated by knowledgeable persons who understand the blockchain network and the processing power required to modify it, it is a potential problem specific to digital currency.

15. **Consensus Algorithms:**

A consensus algorithm is a process in CSE (computer science and engineering) used to attain agreement on a single data value among distributed processes. Agreement-based methods are used in networks with several faulty nodes to achieve accuracy. In distributed computing and multiagent systems, resolving the "consensus problem" is critical. Consensus algorithms in blockchains ensure that each new block contributed to the network is the sole version of the truth that all nodes in a distributed/decentralized computing network agree on.

16. **PROOF OF WORK (PoW):**

The original consensus algorithm in a blockchain network is Proof of Work (PoW). The algorithm authenticates the transaction and adds a new block to the chain. A bunch of users compete against each other to complete the network transaction in this method. Mining is the process of competing against one another. Bitcoin is the most well-known PoW application.

17. **PROOF OF STAKE (PoS):**

Proof of stake (PoS) protocols are a kind of blockchain consensus method that determines validators based on their cryptocurrency holdings. PoS systems, unlike proof of work (PoW) protocols, do not encourage excessive energy consumption. Peercoin was the first cryptocurrency to implement PoS in 2012. Cardano is the most valuable proof-of-stake blockchain in terms of market capitalization.

18. **Practical Byzantine Fault Tolerance(PBFT)**

The Byzantine Fault Tolerance is a consensus algorithm for blockchain. Asynchronous systems were created with PBFT in the account. It is designed to have a reduced overhead time. Its purpose was to address a number of issues with existing Byzantine Fault Tolerance methods. Distributed computing and blockchain are two examples of application areas. It initiates the PBFT consensus network's three phases: pre-prepare, prepare, and commit. The header node multicasts a pre-prepared message to other nodes in the PBFT network during the pre-prepare phase.

19. **Delegated Byzantine Fault Tolerance (DBFT) Algorithm:**

DBFT is a consensus technique that many blockchains and cryptocurrency adopters are unfamiliar with. NEO, dubbed the "Ethereum of China," launched the Delegated Byzantine Fault Tolerance consensus. By digitizing assets and establishing smart contracts on the blockchain, this Chinese blockchain aims to create a "smart economy." DBFT's voting approach, similar to Delegated Proof-of-Stake consensus, enables large-scale involvement, according to its inventors. Absolute finality is one of the main advantages of adopting the DBFT process.

A block cannot be bifurcated after final confirmation; hence the transaction cannot be cancelled or rolled back. However, this is a two-sided weapon. The fact that NEO is not a completely decentralized network ensures the network's finality. Despite NEO's efforts in this area, the blockchain currently has only seven nodes and a few delegates in operation. The majority are linked to the NEO council.

20. **IoT:**
    Pharmaceutical companies are progressively ready to explore with IoT technology to boost reliability, improve productivity, reduce overall production errors, and raise stakeholder expectations for medicine efficacy. The use of IoT in pharmaceutical manufacturing, distribution network (supply chain), clinical development, and patient engagement not only helps to reduce the time to market for medications, but it also aids in the discovery of flaws throughout the value chain via real-time data feed, which aids in regulatory compliance.

21. **FIAT Money:**
    FIAT money is a government-issued currency that authorizes central banks to regulate the amount of money generated, giving more power over the economy. The most of paper currencies, including Indian Currency and U.S dollar, are fiat currencies.

22. **RFID:**
    RFID tags are a type of stalk system that uses radiofrequency to recognize, detect, track, and communicate with products and users. RFID tags are smart labels that can carry various information, which include serial numbers, a brief explanation, and even an entire page containing whole information.

# References

1. Ahokangas P, Matinmikko-Blue M, Yrjola S, Seppänen V, Hämmäinen H, Jurva R, Latva-aho M (2019) Business models for local 5G micro operators. IEEE Trans Cogn Commun Netw. PP. 1–1. https://doi.org/10.1109/TCCN.2019.2902547
2. Chowdhury MZ, Shahjalal M, Ahmed S, Jang YM (2019) 6G wireless communication systems: Applications, requirements, technologies, challenges, and research directions. arXiv:1909. 11315. http://arxiv.org/abs/1909.11315
3. Tariq F, Khandaker M, Wong KK, Imran M, Bennis M, Debbah M (2019) A speculative study on 6G. arXiv:1902.06700. http://arxiv.org/abs/1902.06700
4. Mettler M (2016) Blockchain technology in healthcare: The revolution starts here. In: IEEE 18th International conference on e-Health networking. applications and services (Healthcom), pp. 1–3, https://doi.org/10.1109/HealthCom.2016.7749510
5. Khatoon A (2020) A Blockchain-based smart contract system for healthcare management. Electronics. 9(94):2169–3536. https://doi.org/10.3390/electronics9010094 www.mdpi.com/journal/electro
6. Singhal B, Dhameja G, Panda (2018) P. S.: How blockchain works, In: Beginning blockchain. Springer, pp. 31–148
7. Li X, Jiang P (2017) A survey on the security of blockchain systems. Future Gen Comput Syst https://doi.org/10.1016/j.future.2017.08.020. ,http://www.sciencedirect.com/science/article/pii/ S0167739X17318332

8. Plotnikov V, Kuznetsova V (2018) The prospects for the use of digital technology "Blockchain" in the pharmaceutical market. MATEC Web of Conferences 193:02029. https://doi.org/10. 1051/matecconf/201819302029

9. Li Z, VatankhahBarenji A, Huang GQ (2018) Toward a blockchain cloud manufacturing system as a peer to peer distributed network platform. Robot Comput Integr Manuf 54:133–144. https:// doi.org/10.1016/j.rcim.2018.05.011

10. Jaag C, Bach C (2017) Blockchain technology and cryptocurrencies: Opportunities for postal financial services. The changing postal and delivery sector. Springer, Cham, pp 205–221

11. Jeppsson A, Olsson O (2017) Blockchains as a solution for traceability and transparency. Retrieved from1–102. https://lup.lub.lu.se/student-papers/search/publication/ 8919957

12. Yli-Huumo J, Ko D, Sujin C, Park S (2016) Where is current research on Blockchain technology?—a systematic review. PLOS ONE | DOI:https://doi.org/10.1371/journal.pone.016 3477

13. Jeff G (2015) Public versus private Blockchains. Part 1: Permissioned Blockchains

14. Jeff G (2015) Public versus private Blockchain. Part 2: PermissionlessBlockchains

15. Du M, Chen Q, Chen J, Ma X (2021) An optimized consortium Blockchain for medical information sharing IEEE Trans Eng Manag. 68(6):1677–1689. https://doi.org/10.1109/TEM.2020. 2966832

16. Jiang S, Wu H, Wang L (2019) Patients-controlled secure and privacy-preserving EHRs sharing scheme based on consortium Blockchain. IEEE global communications conference (GLOBECOM). pp. 1–6. https://doi.org/10.1109/GLOBECOM38437.2019.9013220

17. Gupta N (2020) a deep dive into security and privacy issues of Blockchain technologies. Handbook of RESEARCH ON BLOCKCHAIN TEChnology. https://doi.org/10.1016/B978-0-12-819816-2.00004-6

18. Reghunadhan R (2020) Ethical considerations and issues of Blockchain technology-based systems in war zones: A case study approach. Handbook of research on Blockchain technology. https://doi.org/10.1016/B978-0-12-819816-2.00001-0

19. Fernando E, Surjandy M (2019), Success factor of implementation Blockchain technology in pharmaceutical industry: a literature review. In: 6th international conference on information technology, computer and electrical engineering (ICITACEE), pp. 1–5. https://doi.org/10.1109/ ICITACEE.2019.8904335

20. Randhir K, Rakesh T (2019) Traceability of counterfeit medicine supply chain through Blockchain. In: 11th international conference on communication system & Networks

21. Kshetri N (2017) Blockchain's roles in strengthening cybersecurity and protecting privacy. Telecommunications Policy 41(10):1027–1038. https://doi.org/10.1016/j.telpol.2017.09.003

22. Tien D, Rui L, Meihui Z, Gang C, Beng O, Ji W (2017). Untangling Blockchain: a data processing view of Blockchain systems. IEEE Trans Knowl Data Engineering https://doi.org/ 10.1109/TKDE.2017.2781227

23. Hewa T, Gür G, Kalla A, Ylianttila M, An, Braeken, Liyanage M (2020) The role of Blockchain in 6G: challenges. Opportunities and research directions. https://doi.org/10.1109/6GSUMM IT49458.2020.9083784

24. Orsenigo L, Pammolli F, Riccaboni M (2001) Technological change and network dynamics: Lessons from the pharmaceutical industry. Research Policy. 30(3):485–508, ISSN 0048–7333, https://doi.org/10.1016/S0048-7333(00)00094-9

25. The Golem Project Crowdfunding Whitepaper (2018)

26. SONM Decentralized Fog Computing Platform (2018) https://docs:sonm:com/

27. Katuwal GJ, Pandey S, Hennessey M, Lamichhane B (2018), A Preprint, arXiv:1812.02776v1 6

28. Kadam A, Patil S, Patil S, Tumkur A (2016) Pharmaceutical waste management an overview. Indian J Pharm Pract 9(1)

29. Dods V, Taylor B (2021) A proposal for decentralized, global, verifiable health care credential standards grounded in pharmaceutical authorized trading partners. Blockchain in healthcare Today, 4. https://doi.org/10.30953/bhty.v4.175

30. Haq I, MuselemuEsuka O (2018) Blockchain technology in pharmaceutical industry to prevent counterfeit drugs. Int J Comput Appl 180(25):0975 – 8887
31. Bamakan SMH, ShimaGhasemzadehMoghaddam SajedehDehghanManshadi (2021) Blockchain-enabled pharmaceutical cold chain : Applications, key challenges, and future trends. J Clean Prod 0959–6526. https://doi.org/10.1016/j.jclepro.2021.127021
32. Mackey TK, Kuo TT, Gummadi B, Clauson KA, Church G, Grishin D, Obbad K, Barkovich R, Palombini M (2019) 'Fit-for-purpose?' – challenges and opportunities for applications of blockchain technology in the future of healthcare. Mackey et al. BMC medicine 17:68 https://doi.org/10.1186/s12916-019-1296-7
33. Aich S, Chakraborty S, Sain M, Lee HI, Kim HC (2019) A review on benefits of IoT integrated Blockchain based supply chain management implementations across different sectors with case study. In: International conference on advanced communications technology(ICACT) ISBN 979–11–88428–02–1 ICACT2019 February 17 ~ 20
34. Lu Y (2020) Security in 6G: the prospects and the relevant technologies. J Ind Integr Manage 5:1–24
35. Carter CR, Rogers DS, Choi TY (2015) Toward the theory of the supply chain,. J Supply Chain Manag 51(2):89–97. https://doi.org/10.1111/jscm.12073
36. Kumar R, Tripathi R (2020) Blockchain-based framework for data storage In Peer-To-Peer scheme using interplanetary file system. Handbook of research on Blockchain technology. https://doi.org/10.1016/B978-0-12-819816-2.00002-2
37. Hirtan L, Krawiec P, Dobre C, Batalla JM (2019) Blockchain-based approach for e-health data access management with privacy protection. In: Proceedings of IEEE 24th international workshop on computer aided modeling and design of communication links and networks (CAMAD),1–7. https://doi.org/10.1109/CAMAD.2019.8858469
38. Fu Y, Zhu J (2019) Big production enterprise supply chain endogenous risk management based on Blockchain. Special section on healthcare information technology for the extreme and remote environments 7:15310
39. Bartlett PA, Julien DM, Baines, (2007) Improving supply chain performance through improved visibility. Int J Logist Manag 18(2):294–313. https://doi.org/10.1108/09574090710816986
40. Taylor D (2016) The pharmaceutical industry and the future of drug development. In Environmental science and technology No. 41 pharmaceuticals in the environment 2016
41. Jabbar S, Lloyd H, HammoudehM. et al (2021) Blockchain-enabled supply chain: analysis, challenges, and future directions. Multimedia Syst 27:787–806. https://doi.org/10.1007/s00 530-020-00687-0
42. Gordon WJ, Catalini C (2018) Blockchain technology for healthcare: facilitating the transition to patientdriven interoperability. Comput Struct Biotechnol J 16 224–230. https://doi.org/10.1016/j.csbj.2018.06.003
43. Ferver K, Burton B, Jesilow P (2009) The use of claims data in healthcare research. Open Public Health J 2:11–24
44. Benchoufi M, Ravaud P (2017) Blockchain technology for improving clinical research quality. Trials 18 335. https://doi.org/10.1186/s13063-017-2035-z
45. Nguyen T, Tran N, Loven L, Partala J, Kechadi MT, Pirttikangas S (2020) Privacy-aware blockchain innovation for 6G: challenges and opportunities proc. 2nd 6G wireless summit (6G SUMMIT), Levi, Finland, Mar, pp. 1–5
46. Dimiter V (2019) Blockchain applications for healthcare data management. Healthc Inform Res. J 25(1) : 51–56. https://doi.org/10.42,58/hir.2019.25.1.51pISSN 2093–3681
47. Laroiya C, Deepika S, Komalavalli C (2020) Applications of Blockchain technology. Handbook of research on blockchain technology. https://doi.org/10.1016/B978-0-12-819816-2.00009-5
48. Rabah K (2017) Challenges & opportunities for blockchain powered healthcare systems: a review. Mara Res. J. Med. Health Sci. 1(1):45–52
49. Zheng Z, Xie S, Dai H, Chen X, Wang H (2017) An overview of Blockchain technology: architecture, consensus, and future trends. IEEE 6th International congress on big data 978–1–5386–1996–4/17. https://doi.org/10.1109/BigDataCongress.2017.85

50. Benchoufi M, Porcher R, Ravaud P (2017) Blockchain protocols in clinical trials: Transparency and traceability of consent. F1000Research, 6, 66
51. Manogaran G, Rawal BS, Saravanan V (2020) Blockchain based integrated security measure for reliable service delegation in 6G communication environment. Comput Commun 161:1–22

## Additional Reading

52. Jeyabharathi D, Kesavaraja D, Sasireka D (2020) Cloud-based blockchaining for enhanced security. India handbook of research on blockchain technology. https://doi.org/Https://Doi.Org/10.1016/B978-0-12-819816-2.00007-1
53. Shukla RG, ShekharShukla AA (2020) Blockchain-powered smart healthcare system. Handbook of research on blockchain technology. https://doi.org/Https://Doi.Org/10.1016/B978-0-12-819816-2.00010-1
54. Engelhardt MA (2017) Hitching healthcare to the chain: an introduction to Blockchain technology in the healthcare sector, technology innovation management review 7(10)
55. Roxanne E, Lisa DK, George PW (2013) Anti counter feiting in the fashion and luxury sectors: trends and strategies. Anti-counterfeiting – a global guide
56. Tierion - Blockchain. https://tierion.com/. [Accessed 01 10 2021]
57. Blockchain in healthcare. https://www.hyperledger.org/wpcontent/uploads/2016/10/ey-blockc hain-in-health.pdf. [Accessed 23 8 2021]
58. Edge intelligence-6Gflagship, 6G white paper
59. Kantola R (2020) Trust networking for beyond 5G and 6G. In: Proceedings of 2nd 6G wireless summit (6G SUMMIT), Levi, Finland, Mar, pp. 1–6

## Web References

60. https://www.javatpoint.com/blockchain-merkle-tree
61. https://en.m.wikipedia.org/wiki/Cryptographic_hash_function
62. https://en.m.wikipedia.org/wiki/Hash_list
63. https://www.blockchain-council.org/blockchain/blockchain-role-of-p2p-network/
64. https://corporatefinanceinstitute.com/resources/knowledge/other/distributed-ledgers/
65. https://link.springer.com/chapter/10.1007%2F978-3-030-34957-8_9
66. https://scholar.google.co.in/scholar?q=distributed+ledger+technology&hl=en&as_sdt=0&as_vis=1&oi=scholart#d=gs_qabs&u=%23p%3DMB2XlBBiQuIJ
67. https://doi.org/10.1007/s40812-019-00138-6
68. https://doi.org/10.1007/s00181-020-01990-5
69. https://doi.org/10.1007/s11187-019-00286-y
70. https://doi.org/10.1007/s43546-021-00090-5
71. https://www.changehealthcare.com/platform
72. https://internetofthingsagenda.techtarget.com/definition/Internet-of- cThings-IoT
73. https://analyticsindiamag.com/blockchain-consensus-algorithms/
74. https://whatis.techtarget.com/definition/consensus-algorithm
75. https://www.javatpoint.com/blockchain-proof-of-work
76. https://en.wikipedia.org/wiki/Proof_of_stake
77. https://clarle.github.io/yui3/yui/docs/attribute/attribute-rw.html
78. https://www.investopedia.com
79. https://finance.yahoo.com/news/delegated-byzantine-fault-tolerance-dbft-090034374.html

80. https://www.geeksforgeeks.org/practical-byzantine-fault-tolerancepbft/#:~:text=Practical%
    20Byzantine%20Fault%20Tolerance%20is,optimized%20for%20low%20overhead%20time.
81. https://www.investopedia.com/terms/p/private-key.asp
82. https://www.paho.org/en/immunization/cold-chain
83. https://www.google.com/search?q=CRYPTOGRAPHIC+HASH&ei=yvJZYaLlLNK48QP
    1ro64Cw&ved=0ahUKEwii8teH7K7zAhVSXHwKHXWXA7cQ4dUDCA4&uact=5&oq=
    CRYPTOGRAPHIC+HASH&gs_lcp=Cgdnd3Mtd2l6EAMyBQgAEIAEMgUIABCABDIF
    CAAQgAQyBQgAEIAEMgUIABCABDIFCAAQgAQyBQgAEIAEMgUIABCABDIFC
    AAQgAQyBQgAEIAESgQIQRgAUJHHZliT02ZgzddmaABwAngAgAGjAYgBugKSA
    QMwLjKYAQCgAQKgAQHAAQE&sclient=gws-wiz
84. https://doi.org/10.1201/9780429466335/handbook-applied-cryptography-alfred-menezes-
    paul-van-oorschot-scott-vanstone
85. https://cleartax.in/s/view-government-electronic-cash-ledger-guide

Printed in the United States
by Baker & Taylor Publisher Services